Lecture Notes in Statistics

Edited by J. Berger, S. Fienberg, J. Gani, K. Krickeberg, I. Olkin, and B. Singer

Norbert Schmitz

with the assistance of
Günter Duscha
Josef Lübbert
Thomas Meyerthole

Optimal Sequentially Planned Decision Procedures

Springer-Verlag
New York Berlin Heidelberg London Paris
Tokyo Hong Kong Barcelona Budapest

Norbert Schmitz
Institut für Mathematische Statistik
Westfälische Wilhelms-Universität Münster
Einsteinstrasse 62
D-4400 Münster/W.
Germany

Mathematics Subject Classification: 62L10

Library of Congress Cataloging-in-Publication Data
Schmitz, N. (Norbert), 1933-
    Optimal sequentially planned decision procedures / Norbert Schmitz
; with the assistance of Günter Duscha, Josef Lübbert, Thomas
Meyerthole.
        p.   cm. — (Lecture notes in statistics ; 79)
    Includes bibliographical references and index.
    ISBN 0-387-97908-5
    1. Sequential analysis. I. Title. II. Series: Lecture notes in
statistics (Springer-Verlag) ; 79.
    QA279.7.S36 1992
    519.5'4—dc20   92-29504

Printed on acid-free paper.

Camera ready copy provided by the author.
Printed and bound by Edwards Brothers, Inc., Ann Arbor, MI.
Printed in the United States of America.

9 8 7 6 5 4 3 2

ISBN 0-387-97908-5 Springer-Verlag New York Berlin Heidelberg
ISBN 3-540-97908-5 Springer-Verlag Berlin Heidelberg New York

Preface

Learning from experience, making decisions on the basis of the available information, and proceeding step by step to a desired goal are fundamental behavioural qualities of human beings. Nevertheless, it was not until the early 1940's that such a statistical theory – namely Sequential Analysis – was created, which allows us to investigate this kind of behaviour in a precise manner.

A. Wald's famous sequential probability ratio test (SPRT; see example (1.8)) turned out to have an enormous influence on the development of this theory. On the one hand, Wald's fundamental monograph "Sequential Analysis" ([Wa]*) is essentially centered around this test. On the other hand, important properties of the SPRT – e.g. Bayes-optimality, minimax-properties, "uniform" optimality with respect to expected sample sizes – gave rise to the development of a general statistical decision theory. As a consequence, the SPRT's played a dominating role in the further development of sequential analysis and, more generally, in theoretical statistics.

In a sharp contrast to that, the SPRT failed to be accepted by practitioners. There are several reasons for this refusal by applied statisticians – no upper bound for the total sample size, need of better trained staff, risk of manipulations of the data, etc. A further objection (which in fact might be the basic reason behind the other objections) is the following: All optimum properties of the SPRT (Bayes-optimality, minimax-property, Wald-Wolfowitz optimality, etc.) are proven under the explicit or implicit assumption of a linear sampling cost function, i.e. the expected cost of the experiment is proportional to the average sample size. But in most cases it is much more "expensive" to prepare, to carry out and to evaluate k single experiments instead of one experiment of size k – e.g. in pharmaceutical analyses calibrating the instruments causes, in general, high costs, in all agricultural investigations one-at-a-time sampling would lead to an absurd loss of time. For a sensible evaluation of statistical procedures one needs, therefore, more realistic assumptions on the sampling cost function.

In order to get some idea about the severe consequences caused already by (additional) fixed costs for each single sample, consider a simple example (for details comp. Chapter III): Let $X_1, X_2, \ldots$ be independent, identically $\mathcal{B}(1, p)$ (binomial-) distributed random variables; and let $p_1 \simeq 0.90^\dagger$, $p_2 = 0.95$. Corresponding to these values and to error bounds $\alpha_1 = \alpha_2 = 0.10$ we consider the SPRT $\delta_{1/9, 9}$ given by the lower bound $k_1 = 1/9$ and the upper bound $k_2 = 9$. The OC-function and the ASN function of this test are shown in figures 0.1 and 0.2 resp.

---

*Terms in brackets [·] refer to the bibliography in Appendix C.

$\dagger$Due to computational reasons we choose $p_1 = 0.90016837105$ (see p. 10)

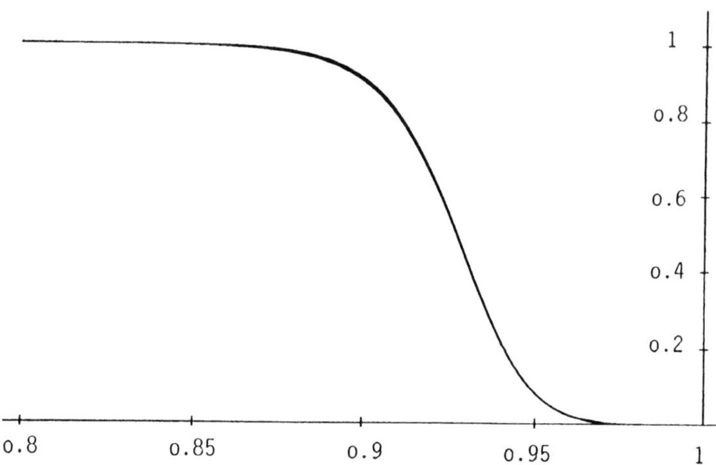

Figure 0.1: OC-function of the SPRT $\delta_{1/9,9}$
(and of the binomial test $\tilde{\delta}$ with fixed sample size 199).

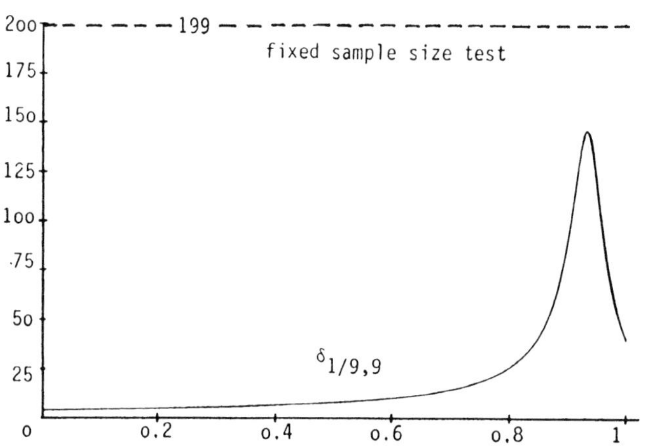

Figure 0.2: Expected sample number of $\delta_{1/9,9}$
and of the fixed sample size test $\tilde{\delta}$

By nature, a fixed-sample procedure involves a single "stage" of sampling when all needed observations are obtained (the need is determined by the error probability). The SPRT, on the other hand, requires successive stages of sampling, each stage obtaining one observation. Consider now the possibility that every stage entails a fixed cost c_0, so that the cost per stage is

$$c(n) = c_0 + n$$

when n observations are obtained at the stage. Then the expected total cost of $\delta_{1/9,9}$ is

(i) for $c_0 = 0$ just the ASN-function (see figure 0.2),

(ii) for $c_0 = 1$ (i.e. fixed costs of the same magnitude as costs per item) shown in figure 0.3.,

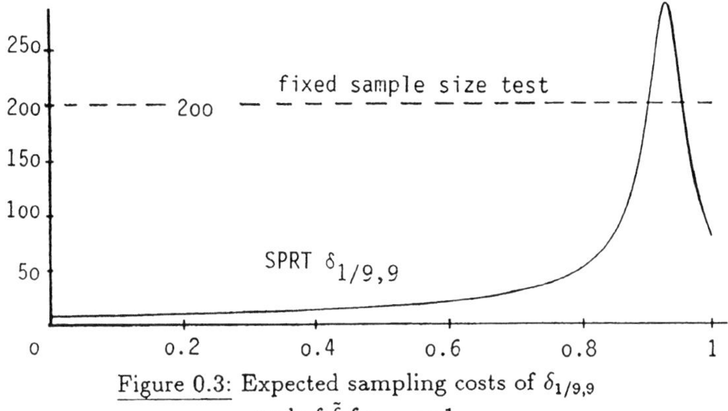

Figure 0.3: Expected sampling costs of $\delta_{1/9,9}$
and of $\tilde{\delta}$ for $c_0 = 1$

(iii) for $c_0 = 20$ (i.e. extremely high fixed costs in comparison to costs per item; occurring e.g. in agricultural experiments) shown in figure 0.4.

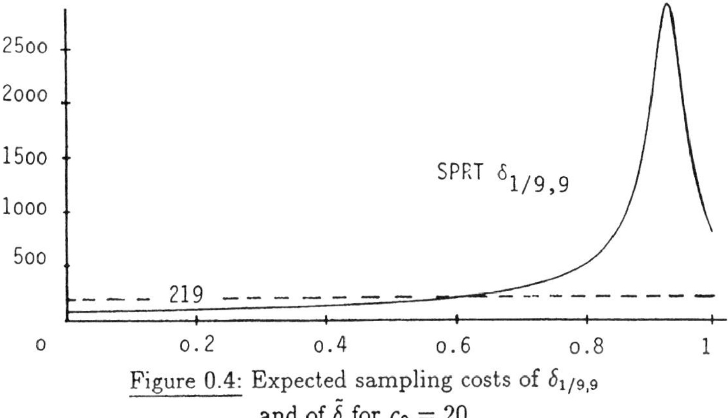

Figure 0.4: Expected sampling costs of $\delta_{1/9,9}$
and of $\tilde{\delta}$ for $c_0 = 20$

This SPRT is compared with the fixed sample size test $\tilde{\delta}$ which decides on the basis of 199 observations (this is the minimal fixed (nonrandomized) sample size such that the error probabilities at p_1 and p_2 keep the values of the SPRT $\delta_{1/9,9}$). The OC-function of $\delta_{1/9,9}$ and $\tilde{\delta}$ nearly coincide (in a graphical representation it is hard to detect any differences); the (expected) sampling costs of $\tilde{\delta}$ are given in figures 0.2-0.4 resp. Already for $c_0 = 1$ – and much more significantly for $c_0 = 20$ – the advantage of a considerably smaller ASN-function of $\delta_{1/9,9}$ is, for the most "important" parameter region between 0.9 and 0.95, more than compensated by the effect of the fixed costs. Obviously, this is an important reason for applied statisticians to refuse the SPRT.

On the other hand, these arguments/objections immediately lead to the question whether it is possible to retain, by suitable modifications, the advantages of sequential procedures without putting up with the disadvantages. Indeed, the aim of these lecture notes is to develop a general theory of group sequential procedures with variable group sizes (*sequentially planned decision procedures*) which yields reasonable compromises for this situation.

In order to demonstrate what can be reached in this way, we again consider our simple example of $\mathcal{B}(1,p)$-random variables: In figures 0.5-0.7 the expected sampling costs of two sequentially planned probability ratio tests (SPPRT; for details see Ch. III), whose OC-functions nearly coincide with that of $\delta_{1/9,9}$ (and $\tilde{\delta}$), are given.

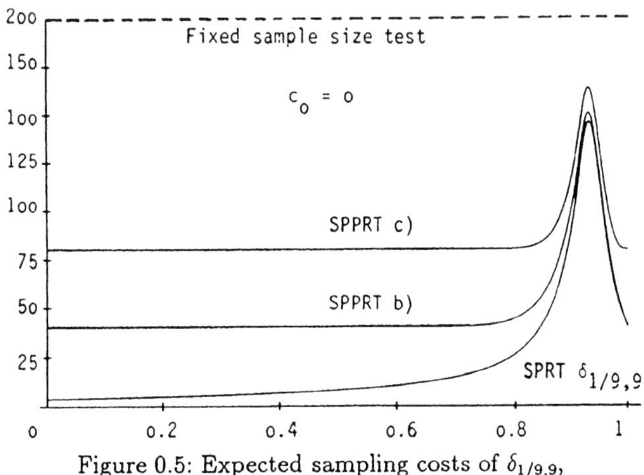

Figure 0.5: Expected sampling costs of $\delta_{1/9,9}$, $\tilde{\delta}$ and two SPPRT's for $c_0 = 0$

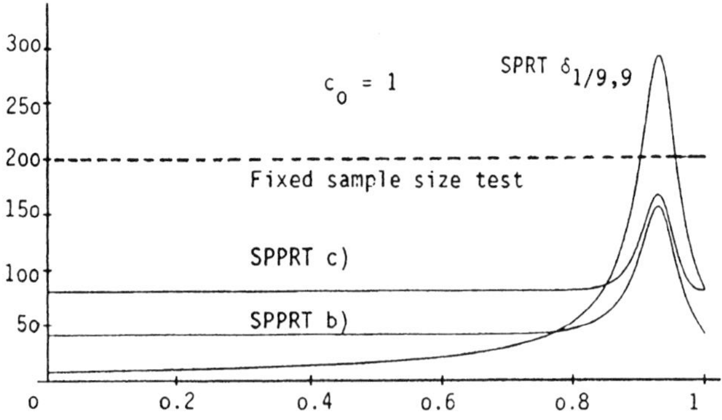

Figure 0.6: Expected sampling costs of $\delta_{1/9,9}$, $\tilde{\delta}$ and two SPPRT's for $c_0 = 1$

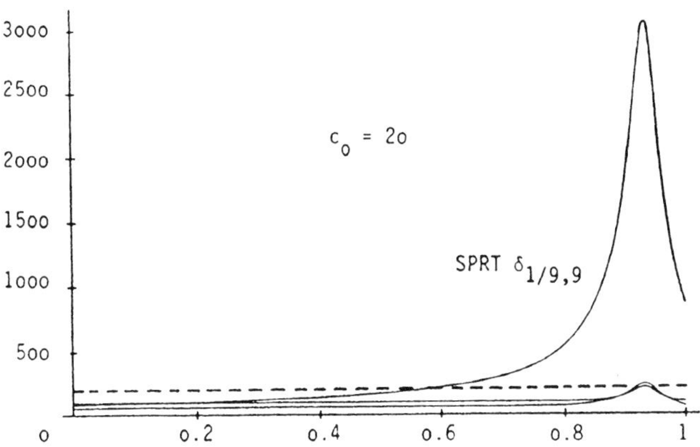

Figure 0.7: Expected sampling costs of $\delta_{1/9,9}$,
$\bar{\delta}$ and two SPPRT's for $c_0 = 20$

These results show that the SPPRT's lead – at least for this special case – to a considerable improvement on the SPRT and, simultaneously, preserve the advantages of sequential methods as compared with fixed sample procedures. In particular, these sequentially planned procedures turn out to be reasonable candidates for *monitoring strategies* for clinical trials (for this aspect comp. also the review article by Jennison and Turnbull [J/T]).

I am very grateful to the *Deutsche Forschungsgemeinschaft*, whose grant No. Schm 677/1-1 gave a number of graduate students (Günter Duscha, Marion Harenbrock, Dietmar Kohlruss, Josef Lübbert, Thomas Meyerthole and Michael Pfannkuche-Winkler) the possibility to work on this project. Many contributions of these gifted young statisticians are incorporated into these lecture notes. Sincere thanks also to the *Volkswagen-Stiftung*, whose Akademie-Stipendium made an intensive cooperation with my students possible.

Last but not least, the production of this work would have been impossible without the valuable assistance of Martina Forstmann, who became a TeX-expert during typing and retyping of several drafts of this volume.

Contents

I. Introduction

§ 1 Sequential statistical procedures

In "classical" mathematical statistics one assumes that a statistical decision (e.g. point estimator for a parameter, test of hypotheses, curve estimator, confidence interval for a parameter, etc.) has to be made on the basis of a *fixed number n* of observed values $x_1, \dots, x_n$ – large parts of the corresponding theory may be found in the books by Lehmann [Le], Bickel/Doksum [B/D] or Witting [Wi1], [Wi2]. For many practical applications this is just the description of the situation the statistician is confronted with: After the observations have been made the investigator shows up with his data and asks for advice on how to analyze them.

It seems obvious that advantageous possibilities are given away by not making use of the information gathered during the course of the investigation. We will illustrate this by a well-known example from statistical quality control.

(1.1) Example ("curtailed inspection")

Batches of items are subjected to an acceptance inspection. According to conditions on the producers and the consumers risk the following inspection plan is used: A batch is rejected if a sample of 20 items contains more than 2 defective; otherwise the batch is accepted ((20,2)-sampling plan).

But observing the special sample

$$\text{E E D E E. D E E }\underline{\text{D}}\text{ E. E E E E D. E E E E E}$$

(E $\sim$ "effective", D $\sim$ "defective") the ninth value already determines the final decision (the batch is rejected). Thus it is possible, without changing the accuracy of the statistical procedure, to save the time and costs of inspecting the additional 11 items. An obvious idea is, therefore, to *curtail* the (20,2)-sampling plan by stopping the inspection as soon as either 3 defective items are found (then the decision is "rejection") or 18 effective items are observed (then the batch is accepted). $\square$

More generally, for an (n, a)-sampling plan (n denoting the sample size, a the acceptance number) the following simple modification seems to be reasonable: Inspect items successively until either

> *the number of defective items is $a+1$*

or

> *the number of effective items is $n-a$.*

Then the statistical decisions exactly coincide with those of the original (n, a)-sampling plan, but in many cases these decisions are made with a smaller number of inspections and, therefore, lower costs. A graphical representation of this *curtailed inspection plan* $(\widehat{n, a})$ can be given in the following way:

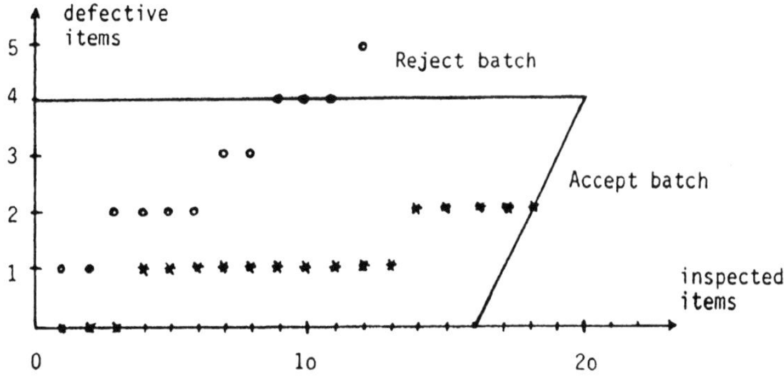

Figure 1.1: Curtailed inspection plan $(\widehat{20,4})$

[The sample path marked by o leads to a "rejection" having inspected 12 items, for the path marked by $\star$ the batch is accepted on the basis of 18 inspections]

Some properties of these curtailed inspection plans are described in section 1.3 of the monograph [We] by Wetherill (see also (1.7)). The main difference between the curtailed inspection plan $(\widehat{n,a})$ and the original plan (n, a) is that the sample size is no longer fixed but depends on the outcome of the successive observations. Therefore, the sample size of a curtailed inspection plan itself is a *random variable* N with the property

$$N \leq n$$

and an average sample number (ASN)

$$E(N) < n$$

for "all" possible distributions of the sample. Depending on the special sample the *sequential* proceeding leads to a more or less considerable saving of observations without any loss of statistical precision.

This simple example already shows an essential general aspect of sequential statistical procedures: *The sample size is not fixed in advance but depends, by successively collecting and evaluating the data, on the information contained in the observations at each stage* – if the data are very informative one will come to a final decision with a small sample, if the observations fail to give clear indications one needs a long investigation to arrive at reliable statistical decisions [1].

§ 2 Objectives of sequential analysis

Example (1.1) moreover gives some hints on possible advantages of sequential methods and on goals a theory of sequential statistical procedures (*sequential analysis*) should

---

[1] Proceeding in this way the statistician at the same time reaches a new "position" within the statistical investigation: While in classical statistics he often has to try to make the best from data which have been collected without consulting him, he now has influence on the pattern (the sampling rules and the final sample size) of the experiment.

try to reach:

(1.2) Saving of observations/observation costs

Curtailed inspection plans $(\widehat{n,a})$ *never* demand a larger number of observations than the corresponding (n,a)-plans. But if one admits random sample sizes the question arises whether one should accept, for certain sample paths, larger numbers of observations if this leads to saving *on the average*. Among others this consideration was a reason for measuring the "demand of sampling" in sequential analysis by the average sample number (ASN) $E_P(N)$ (where P denotes the "true" distribution) (objections against this concept will be discussed in § 5). Then one tries to minimize, under all statistical procedures with certain accuracy properties "the" ASN. It turns out that, in many cases, considerable improvements upon classical fixed sample procedures are available (see e.g. the book [Gh] by B.K. Ghosh and pages 12/13). $\square$

This objective of saving observations/costs is closely connected with a goal of sequential analysis which is, due to ethical reasons, of special importance in biometrical problems:

(1.3) Early use of better treatments

Again, this aspect can easily be explained by a simple example:

(1.4) Example

Two "competing" drugs A and B – e.g. a traditional medicine and a newly developed drug – are to be compared in a clinical study, i.e. by treating patients with these drugs. If during the course of the investigation clear indications are obtained for a considerable superiority of the new medicine B it can, for ethical reasons, hardly be justified to treat further patients with the inferior drug A or to continue a treatment with A – instead one has to give up the original design of the experiment. Moreover, if during the clinical study the (serious) suspicion arises that the new medicine might have dangerous side effects it cannot be justified to use the drug until a *fixed* number of comparisons have been made. In this case, a reasonable objective is to keep the proportion of patients treated with the inferior drug as small as possible. $\square$

The reasons for using sequential statistical methods, mentioned in (1.2) and (1.3), concern situations where fixed sample procedures should be *improved* . Moreover, there are statistical problems which can be solved only by sequential methods:

(1.5) Solving problems which cannot be handled by fixed sample procedures

To illustrate this aspect we again consider a – surprisingly simple – example:

(1.6) Example

In many statistical problems (e.g. simulation studies, experimental comparisons, econometrical analyses) it is assumed – refering to the central limit theorem – that the observed data

$$x = (x_1, \ldots, x_n)$$

4

are realizations of independent $\mathcal{N}(\mu, \sigma^2)$-(normal) distributed random variables; a confidence interval for the unknown mean μ has to be constructed. A reasonable accuracy requirement is that, on the one hand, a given level $1 - \alpha$ is guaranteed (i.e. attention is restricted to level $1 - \alpha$ confidence intervals) and that, on the other hand, the statement is "precise" enough in the sense that the interval has at most the length $2\,l$ (where l is given).

While in the case of variance σ^2 known this requirement can be fulfilled by simply taking

$$C_n(x) = [\overline{x}_{(n)} - l; \overline{x}_{(n)} + l]$$

where [2] $\overline{x}_{(n)} := \sum_{i=1}^{n} x_i/n$, $n := \lceil \sigma^2 u_{\alpha/2}^2/l \rceil$, there does not exist in the case of σ^2 unknown (which underlies most applications) any confidence interval with fixed sample size which fulfills this requirement (theorem of Dantzig, see [Da]). But there exist sequential confidence intervals – e.g. Stein's two-stage procedure (see (1.10)) – with the desired properties. $\square$

§ 3 Historical remarks on the development of sequential analysis

The basic idea of sequential analysis to use, at any time point, all available information for the decisions on the future actions is, of course, a general basis of rational behaviour. Nevertheless the (mathematical) formulation and systematic use of this idea in statistics are relatively young:

The first formalized sequential procedure seems to be the "double sample" inspection plan by H.F. Dodge and H.G. Romig (1929)[3]. Here the results of a first sample may yield enough information to make the second sample superfluous. This inspection plan was generalized by W. Bartky (1943)[4] to a multi-stage procedure. Although there were some other proposals of multi-stage procedures (e.g. by Mahalanobis (1940) and by Hotelling (1943)) a systematic development of sequential methods did not start until the last two years of World War II – and that independently in USA and Great Britain.

Milton Friedman and *Allen Wallis*, members of the *"Statistical Research Group"* at Columbia University[5], took up the suggestion of Captain G.L. Schnyler to make tests for the comparison of two proportions more economical by proceeding sequentially. After some preliminary considerations they posed the problem in March/April 1943 to *Abraham Wald* [6]. He succeeded in developing a general theory of sequential analysis; in particular he proposed the *"Sequential Probability Ratio Test"* (SPRT; see (1.8)). For reasons of (military) security the results of the Statistical Research Group were

---

[2] $\lceil y \rceil$ denotes the smallest integer $\geq y$ and $u_{\alpha/2}$ is the $\alpha/2$-quantile of the $\mathcal{N}(0,1)$-distribution.

[3] Dodge, H.F./ Romig, H.G.: A method of sampling inspection. Bull. Syst. Techn. J. 8(1929), 613-631

[4] Bartky, W.: Multiple sampling with constant probability. Ann. Math. Statist. 14(1943), 363-377

[5] For an extended description of the work of that group we refer to the article "The Statistical Research Group, 1942-1945" by W.A. Wallis; J. Amer. Statist. Assoc. 75 (1980), 320-335

[6] Abraham Wald, born Oct. 31. 1902 in Cluj (Romania), studied in Vienna, emigrated in 1938 to USA, died Dec. 13. 1950 by an aeroplane crash in India. He made important contributions to geometry, mathematical economics and to statistical decision theory.

not allowed to be published until 1945[7]; but they were collected for a *Final Report* submitted to the Applied Mathematics Panel, National Defense Research Committee.[8] Therefore, when suspended from keeping the results secret, A. Wald was able to publish in a voluminous paper ("Sequential tests of statistical hypotheses", Ann. Math. Statist. 16(1945), 117-186) a variety of important results; soon after that the book "Sequential Analysis" (1947) by A. Wald appeared – sequential analysis is one of the very few theories which almost immediately started with a monograph (later on this fact turned out to be quite important because the points had been shifted). Moreover, the results of A. Wald and the other members of the Statistical Research Group were soon accompanied by a series of further developments e.g. the assertions on Stein's two-stage procedure ([St], see example (1.10)).

Completely independent from the developments in USA in the same year of 1943 systematic investigation of sequential methods was started in Great Britain[9]. Here also relations were pointed out to classical questions of ruin probabilities (in the discrete case) and to diffusion problems (in the continuous case), but the basic objective[10] – namely the saving of sampling costs - was the same as that of the Statistical Research Group. It was again security requirements that caused the results to remain unpublished until 1945.

In consideration of the variety of new results many statisticians may have got the impression that sequential analysis was the philosopher's stone for mathematical statistics. This feeling was, moreover, intensified by Wald and Wolfowitz's proof of the simultaneous optimality of the sequential probability ratio tests[11] and further optimality properties (in the Bayesian sense) arising in this connection, by more general optimality assertions on sequential methods shown by Arrow, Blackwell and Girshick[12], and by the fact that A. Wald was inspired by these results to develop a general statistical decision theory (see the monograph "Statistical Decision Functions" (1950) by A. Wald).

During the 50's this euphoria was followed by a certain disillusionment: One problem (recognized already by A. Wald) is that the (implicit) assumption of linear sampling costs is far from being realistic in many applications and that (therefore) the average sample number fails to be a suitable criterion for evaluating statistical procedures (for this aspect see also § 5). Moreover, it turned out that sequential tests which have excellent properties for distinguishing between simple hypotheses – as e.g. Wald's sequential probability ratio tests – may prove to be poor when testing composite hypotheses. This

---

[7]For the anecdote that Wald had as enemy alien no access to his own research results comp. the article by W.A. Wallis (footnote 5).

[8]Statistical Research Group, Columbia University (1945): Final Report (submitted to the Applied Mathematics Panel, National Defense Research Committee in completion of research under Contract OEMsr-618 between the trustees of Columbia University and the Office of Scientific Research and Development).

[9]We refer to the report "Sequential tests in industrial statistics", J. Royal Statist. Soc., Supplement 8 (1946), 1-26 (with discussion) by G.A. Barnard.

[10]Barnard, G.A.: Economy in sampling with special reference to engineering experimentation. Brit. Min. Supply Adv. Serv., Stat. Math. and Qual. Control. Techn. Report QC/R/7,I (1944).

[11]Wald, A./Wolfowitz, J.: Optimum character of the sequential probability ratio test. Ann. Math. Statist. 19(1948), 326-339.

[12]Arrow, K.I./Blackwell, D./Girshick, M.A.: Bayes and minimax solutions of sequential decision problems. Econometrica 17(1949), 213-244.

6

disappointment that (too) high expectations were not fulfilled led to a certain neglect of sequential analysis.

A new development started with a paper of T.W. Anderson whose title "A modification of the SPRT to reduce the sample size" sounds, in consideration of the Wald-Wolfowitz optimality result, somewhat challenging. Since then statisticians have been exploiting the advantages of sequential methods without purchasing the disadvantages (e.g. large variations of observation times) caused by laying too much emphasis on one special optimality criterion. One is no longer interested in statistical procedures which are optimal with respect to a single criterion but looks for procedures which are reasonable from different perspectives.

§ 4   Examples of sequential procedures; purely sequential statistical decision procedures

To illustrate how sequential procedures work we give some simple examples which will be used in the sequel several times:

(1.7) Example *(curtailed inspection; see (1.1))*

Let $X_1, \ldots, X_n$ be independent, $\mathcal{B}(1, p)$ (binomial-) distributed random variables (with the meaning $P_p(X_i = 1) = p = 1 - P_p(X_i = 0)$ and

$$
X_i = \begin{cases} 1 & \text{defective} \\ & \sim \text{the i-th item is} \qquad\qquad) \\ 0 & \text{effective} \end{cases}
$$

and let $S_m := \sum_{i=1}^m X_i$ (i.e. the number of defective items among the first m inspected), $1 \leq m \leq n$. The curtailed inspection plan $(\widehat{n, a})$ (see (1.1)) is now defined by *stopping* the inspection at time point

$$
N := \min\{m \leq n : S_m = a + 1 \text{ or } S_m = m + a - n\}
$$

and making the *final decision*

$$
\begin{array}{ccc}
\text{Acceptance} & & N + a - n \\
& \text{for } S_N = & \\
\text{Rejection} & & a + 1
\end{array} \quad .
$$

The probability of rejecting the batch (when p is the true parameter) is

$$
\begin{aligned}
P_p(S_N = a + 1) &= \sum_{m=a+1}^{n} P_p(S_m = a + 1, N = m) \\
&= p^{a+1} \sum_{m=a+1}^{n} \binom{m-1}{a}(1 - p)^{m-a-1} \\
&= P_p(S_n \geq a + 1) = \sum_{m=a+1}^{n} \binom{n}{m} p^m (1 - p)^{n-m}
\end{aligned}
$$

(i.e. just the same as for the (n, a)-plan in (1.1)). Similarly, the average sample number is

$$E_p(N) = \sum_{m=1}^{n} m P_p(N = m) =$$

$$= \sum_{m=a+1}^{n} m\, P_p(S_m = a+1, N = m) +$$

$$+ \sum_{m=n-a}^{n} m\, P_p(S_m = m + a - n, N = m)$$

$$= p^{a+1} \sum_{m=a+1}^{n} m \binom{m-1}{a}(1-p)^{m-a+1} +$$

$$+ \sum_{m=n-a}^{n} m \binom{m-1}{n-a-1} p^{m+a-n}(1-p)^{n-a}$$

$$= (a+1)p^{a+1} \sum_{i=1}^{n-a} \binom{i+a}{1+a}(1-p)^{i-1} +$$

$$+ (n-a)(1-p)^{n-a} \sum_{i=0}^{a} \binom{i+n-a}{i} p^{i}.$$

Therefore it makes no sense to speak of *the* (expected) sample size – instead the expected value $E_p(N)$ really depends on the (unknown!) distribution-parameter p.

Figures 1.2 and 1.3 show, for the special curtailed inspection plan $(\widehat{20,2})$, graphical representations of the probabilities $P_p(S_N = 3)$ for "rejection" and the expected sample sizes $E_p(N)$.

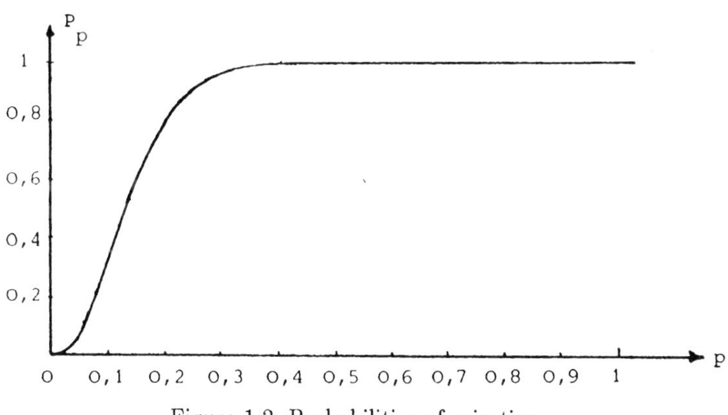

Figure 1.2: Probabilities of rejection
of the $(20,2)$ (and $(\widehat{20,2})$) plan

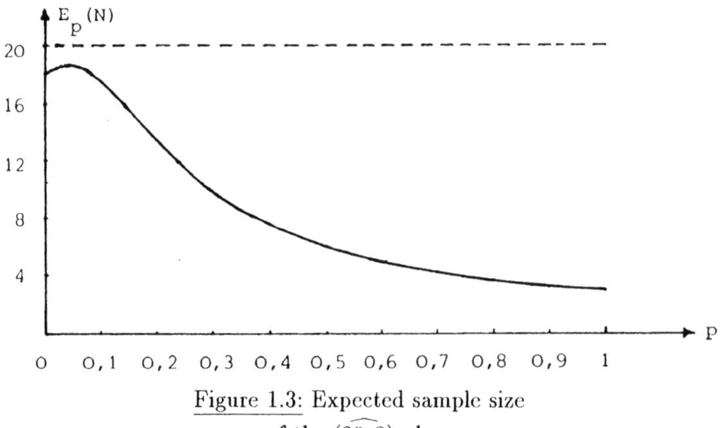

Figure 1.3: Expected sample size
of the $(\widehat{20,2})$ plan.

In particular for higher values of p (i.e. bad production quality) one gains a considerable saving of (expected) sample sizes.

(1.8) Example *(Sequential probability ratio test for independently repeated experiments)*

Let $X_1, X_2, \ldots$ be independent and identically distributed random variables[13] and let

$$H_j : P^{X_1} = Q_j, \ j = 1, 2$$

be two simple hypotheses concerning the distribution of X_1 (where, of course, $Q_1 \neq Q_2$). The basic idea of the sequential probability ratio tests (SPRT) is to continue sampling until the observed data allow a clear distinction between the hypotheses. To make this idea (mathematically) precise let

f_j be densities of Q_j with respect to a dominating measure μ (e.g. $\mu = Q_1 + Q_2$), i.e. $f_j = dQ_j/d\mu, \quad j = 1, 2,$
q be the *likelihood ratio* of f_1 and f_2, i.e.

$$q(x) = \begin{cases} f_2(x)/f_1(x) & f_1(x) > 0, f_2(x) \geq 0 \\ \infty & \text{for} \quad f_1(x) = 0, f_2(x) > 0 \\ 1 & f_1(x) = f_2(x) = 0, \end{cases}$$

$$q_n(x_1, \ldots, x_n) := \prod_{i=1}^n q(x_i), \ n \in \mathbb{N},$$

and

$$k_1, k_2 \in \mathbb{R} \text{ such that } 0 < k_1 \leq 1 \leq k_2 < \infty, k_1 \neq k_2.$$

The decision procedure δ_{k_1, k_2},

- to continue sampling until q_n leaves the interval (k_1, k_2), i.e. until

$$N := \inf\{n \in \mathbb{N} : q_n(X_1, \ldots X_n) \notin (k_1, k_2)\}$$

(where $\inf \emptyset := \infty$), and

---

[13]This situation will be called "iid case".

- to decide in favour of H_1, if $q_N(X_1,\ldots,X_N) \le k_1$, and in favour of H_2, if $q_N(X_1,\ldots,X_N) \ge k_2$, i.e. to choose (analogously to the Neyman-Pearson lemma) the terminal decision

$$\begin{array}{lll} accept\ H_1 & & \le k_1 \\ & if\ \prod_{i=1}^{N} \frac{f_2(x_i)}{f_1(x_i)} & \\ accept\ H_2 & & \ge k_2, \end{array}$$

is called SPRT with stopping bounds k_1 and k_2.

A graphical illustration of this procedure is given by means of two special sample paths in figure 1.4.

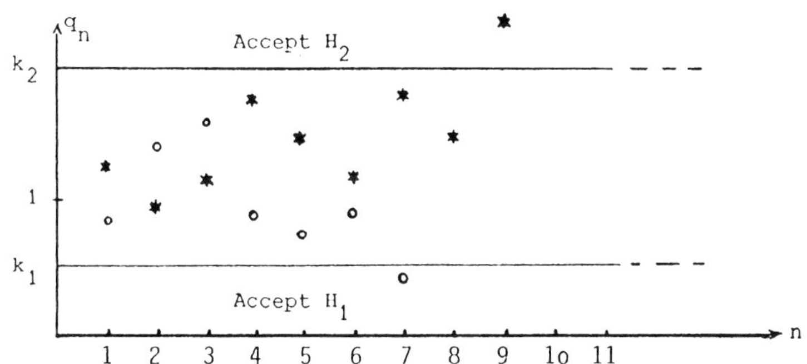

Figure 1.4: SPRT with stopping bounds k_1, k_2
(the sample path marked by $\star$ leads to the acceptance of H_2 by 9 observations,
the path marked by o to the acceptance of H_1 by 7 observations)

The statement "the observed data allow a clear distinction between the hypotheses" is made precise on the basis of the likelihood ratio: $q_n(x_1,\ldots,x_n)$ falling short of k_1 gives reason to accept $H_1, q_n(x_1,\ldots,x_n)$ exceeding k_2 is taken as a hint in favour of H_2.

For SPRT's there are no upper bounds on the sample size – it is possible that $q_n(x_1,\ldots,x_n) \in (k_1,k_2)$ for all $n \in$ IN. Statistical decision procedures which, with positive probability, never reach a terminal decision – and thus have infinite expected sample size – are rarely interesting for practical applications[14]. But it turns out that (at least in the iid-case) every SPRT is finite with probability 1 and that moreover all moments of the sample size – in particular the expected values $E_j(N)$ – are finite. Furthermore, the famous theorem of Wald and Wolfowitz (see e.g. [Gh] or [Ir]) states that every SPRT has, for fixed error probabilities, the smallest expected sample sizes under Q_1 and Q_2.

For the connection between the constants k_i defining the SPRT δ_{k_1,k_2} and its error probabilities $e_j(\delta_{k_1,k_2})$ one obtains the following simple inequalities and approximations.

---

[14]Exceptions arise in "open ended tests" in which infinite samples are accepted for one hypothesis (e.g. for the superior drug).

(1.9) Lemma

For SPRT's in the iid-case we have

a) $\quad e_1(\delta_{k_1,k_2}) \leq (1 - e_2(\delta_{k_1,k_2}))/k_2$

$\qquad e_2(\delta_{k_1,k_2}) \leq (1 - e_1(\delta_{k_1,k_2}))\, k_1.$

b) $e_1(\delta_{\alpha_2,1/\alpha_1}) \leq \alpha_1, \quad e_2(\delta_{\alpha_2,1/\alpha_1}) \leq \alpha_2,$

$\qquad$ *for* $0 < \alpha_1 < 1$ *and* $0 < \alpha_2 < 1.$

c)

$$e_1(\delta_{\hat{k}_1,\hat{k}_2}) \leq \frac{1 - e_2(\delta_{\hat{k}_1,\hat{k}_2})}{1 - \alpha_2}, \quad e_2(\delta_{\hat{k}_1,\hat{k}_2}) \leq \frac{1 - e_1(\delta_{\hat{k}_1,\hat{k}_2})}{1 - \alpha_1}$$

and

$$e_1(\delta_{\hat{k}_1,\hat{k}_2}) + e_2(\delta_{\hat{k}_1,\hat{k}_2}) \leq \alpha_1 + \alpha_2$$

for $0 < \alpha_1 < \alpha_1 + \alpha_2 < 1$ *and* $\hat{k}_1 := \alpha_2/(1 - \alpha_1), \; \hat{k}_2 := (1 - \alpha_2)/\alpha_1.$

Proof (for the first inequalities):

a) $\quad e_1(\delta_{k_1,k_2}) = \sum\limits_{n=1}^{\infty} \int\limits_{\{N=n,q_n>1\}} dQ_1^n = \sum\limits_{n=1}^{\infty} \int\limits_{\{N=n,q_n>1\}} f_1^{(n)} d\mu^n$

$\qquad \leq \sum\limits_{n=1}^{\infty} \int\limits_{\{N=n,q_n>1\}} \frac{1}{k_2} f_2^{(n)} d\mu, \text{ since } q_n = f_2^{(n)}/f_1^{(n)} > k_2 \text{ on } \{\dots\}$

$\qquad = \frac{1}{k_2} \sum\limits_{n=1}^{\infty} \int\limits_{\{N=n,q_n>1\}} dQ_2^n = (1 - e_2(\delta_{k_1,k_2}))/k_2.$

b) From a) follows

$$e_1(\delta_{\alpha_2,1/\alpha_1}) \leq (1 - e_2(\delta_{\alpha_2,1/\alpha_1}))/(1/\alpha_1) \leq \alpha_1.$$

c) The first two inequalities are direct consequences of a); multiplying the first one by $(1 - \alpha_2)$ and the second one by $(1 - \alpha_1)$ and adding them up yields the third inequality. $\qquad\qquad\qquad\qquad\qquad\qquad\qquad\qquad\qquad\qquad\qquad\qquad$ $\square$

Part (1.9) c) gives a very simple and surprisingly precise advice on how to choose constants k_i in order to attain given values α_i approximately as error probabilities. For the special case that P^{X_i} are $\mathcal{B}(1, p_j)$ (binomial-) distributions and that $p_1 \cong 0,90$ and $p_2 = 0,95$ one obtains for example[15]

---

[15]Dr. K.-H. Eger (TH Chemnitz, Germany) has computed these values by using the algorithm contained in his book [Eg]; for computational reasons he used $p_1 = 0,90016837105$.

	$\hat{k}_1$	$\hat{k}_2$	$e_1(\delta_{\hat{k}_1,\hat{k}_2})$	$e_2(\delta_{\hat{k}_1,\hat{k}_2})$
$\alpha_1 = 0,10,\quad \alpha_2 = 0,10$	1/9	9	0,0998	0,0791
$\alpha_2 = 0,05$	1/18	19/2	0,0987	0,0396
$\alpha_2 = 0,01$	1/90	9,9	0,0973	0,0079
$\alpha_1 = 0,05,\quad \alpha_2 = 0,10$	2/19	18	0,0499	0,0791
$\alpha_2 = 0,05$	1/19	19	0,0493	0,0396
$\alpha_2 = 0,01$	1/95	99/5	0,0487	0,0079
$\alpha_1 = 0,01,\quad \alpha_2 = 0,10$	10/99	90	0,0099	0,0787
$\alpha_2 = 0,05$	5/99	95	0,0098	0,0394
$\alpha_2 = 0,01$	1/99	99	0,0098	0,0079

For this special case that the random variables X_i follow a $\mathcal{B}(1,p)$ (binomial-) distribution – i.e. dichotomic variables ("effective – defective", "success – failure") are observed – the SPRT's δ_{k_1,k_2} allow the following representation: For

$$Q_j = \mathcal{B}(1,p_j),\ j = 1,2$$

(where, without loss of generality, $p_1 < p_2$ is assumed) the likelihood ratio after n observations is

$$q_n(x_1,\dots,x_n) = \frac{p_2^s(1-p_2)^{n-s}}{p_1^s(1-p_1)^{n-s}} \text{ where } s = \sum_{i=1}^{n} x_i$$
$$= \left(\frac{p_2(1-p_1)}{p_1(1-p_2)}\right)^s \left(\frac{1-p_2}{1-p_1}\right)^n.$$

Using the SPRT δ_{k_1,k_2} sampling is continued until $q_n(x_1,\dots,x_n) \notin (k_1;k_2)$, i.e. until

$$s \notin \left(\frac{\log\left(k_1\left(\frac{1-p_1}{1-p_2}\right)^n\right)}{\log\left(\frac{p_2(1-p_1)}{p_1(1-p_2)}\right)}; \frac{\log\left(k_2\left(\frac{1-p_1}{1-p_2}\right)^n\right)}{\log\left(\frac{p_2(1-p_1)}{p_1(1-p_2)}\right)}\right)$$

and a decision in favour of H_1 is taken if s is on or below the lower bound and in favour of H_2 if s is on or above the upper bound. Therefore, δ_{k_1,k_2} allows the following graphical representation

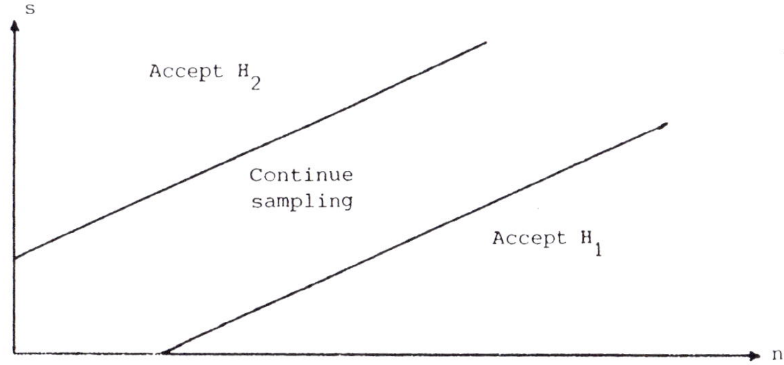

Figure 1.5: SPRT for binomial distributions

For dichotomic variables the SPRT may be compared with the fixed-sample test whose sample size n_0 is minimal in keeping for p_1 and p_2 the same error probabilities as the SPRT. For $p_1 \cong 0,90$ and $p_2 = 0,95$ one obtains e.g. the following values

k_1	k_2	$e_1(\delta_{k_1,k_2})$	$e_2(\delta_{k_1,k_2})$	$E_1(N)$	$E_2(N)$	n_0
1/9	9	0,0998	0,0791	96,46	111,65	199
1/18	9,5	0,0987	0,0396	126,87	124,14	257
1/90	9,9	0,0973	0,0079	198,16	136,46	370
2/19	18	0,0499	0,0791	108,62	149,82	263
1/19	19	0,0493	0,0396	140,72	164,08	322
1/95	19,8	0,0487	0,0079	215,74	177,71	449
10/99	90	0,0099	0,0787	120,62	239,31	399
5/99	95	0,0098	0,0394	153,99	257,45	472
1/99	99	0,0098	0,0079	231,55	273,60	614

$E_i(N)$: Expected sample sizes of δ_{k_1,k_2}; n_0: Minimal sample size of a (fixed sample) binomial test whose error probabilities keep the values $e_i(\delta_{k_1,k_2})$.

This shows that proceeding sequentially leads to a considerable saving of expected sample numbers (a saving of about 50 % when $p = p_1$ or p_2).

Although these SPRT's are, first of all, formulated for the simple hypotheses

$$H_j : P^{X_1} = \mathcal{B}(1,p_j), \ j = 1,2,$$

they may be used in an obvious way also for the composite hypotheses

$$\tilde{H}_1 : P^{X_1} \in \{\mathcal{B}(1,p) : p \leq \tilde{p}_1\}, \ \tilde{H}_2 : P^{X_1} \in \{\mathcal{B}(1,p) : p \geq \tilde{p}_2\}$$

where $p_1 \leq \tilde{p}_1 \leq \tilde{p}_2 \leq p_2$; figure 1.5 illustrates how to proceed.

For the SPRT $\delta_{1/9,9}$ and the standard binomial test one obtains the powers as follows[16]

p	$SPRT\delta_{1/9,9}$	*Binomial test*
0.80	0.00018	0.00000
0.82	0.00059	0.00001
0.84	0.00199	0.00013
0.86	0.00693	0.00165
0.88	0.02551	0.01551
0.90	0.09871	0.09671
0.92	0.35394	0.36742
0.94	0.78818	0.78323
0.96	0.97774	0.98537
0.98	0.99946	0.99999

Table 1.1: Power-function of the SPRT $\delta_{1/9,9}$ and of the binomial test with fixed sample size 199 (the two graphs nearly coincide)

---

[16]All these values have been computed by Dr. Marion Harenbrock (Institut für Mathematische Statistik der Univ. Münster)

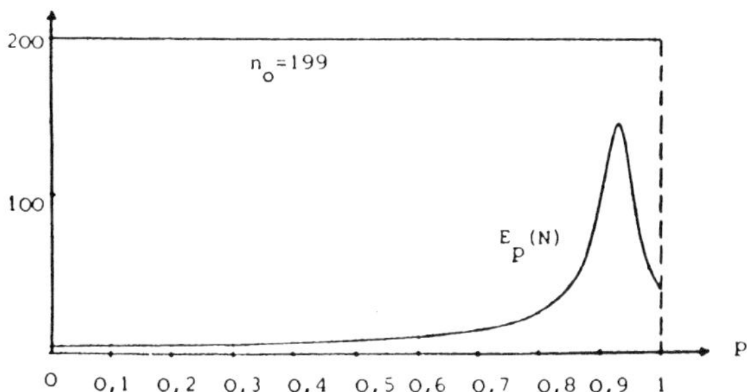

Figure 1.6: Expected sample sizes $E_p(N)$ of the SPRT $\delta_{1/9,9}$ (and of the binomial test)

(1.10) Example *(Stein's two stage procedure; see (1.6))*
Let $X_1, X_2, \ldots$ be independent and identically $\mathcal{N}(\mu, \sigma^2)$-(normally-) distributed random variables whose mean μ and variance σ^2 are unknown. For given values $\ell > 0$ and $\alpha \in (0; 1)$ one looks for a level $1 - \alpha$ confidence interval with fixed width 2ℓ. Whilst there does not exist any fixed sample size confidence interval with these properties (theorem of Dantzig [Da]) this goal can easily be achieved by the following procedure where an initial sample is drawn at first and, on the basis of the gathered information, then the final sample size is determined:

(1.11) Lemma (theorem of Stein [St])

Let $n_1 \geq 2$ and
$$N := \max\{n_1, \lceil S_{n_1}^2 \cdot t_{n_1-1;\alpha/2}^2 / \ell^2 \rceil\}$$

where $t_{n;\gamma}$ denotes the γ-quantile of the central t_n-distribution, $\overline{X}_{(n)}$ the sample mean $\sum_{i=1}^n x_i/n$, and S_n^2 the sample variance $S_n^2 = \sum_{i=1}^n (X_i - \overline{X}_{(n)})^2 / (n-1)$. Then
$$\left[\overline{X}_{(N)} - \ell, \; \overline{X}_{(N)} + \ell\right]$$

yields a (sequential) confidence interval at level $1 - \alpha$ for μ.

Proof: N is a random variable which is determined by $S_{n_1}^2$ (and the constants n_1, ℓ and α). Since $\overline{X}_{(n_1)}$ and $S_{n_1}^2$ are independent (see e.g. [Wi 1], 1 44a)) this holds true also for $\overline{X}_{(n_1)}$ and N; moreover N and $(X_{n_1+1}, \ldots)$ are independent. Therefore

$$P_{\mu,\sigma^2}(|\overline{X}_{(N)} - \mu| \leq \ell) = \sum_{n=n_1}^{\infty} P_{\mu,\sigma^2}(|\overline{X}_{(N)} - \mu| \leq \ell, \; N = n)$$

$$= \sum_{n=n_1}^{\infty} P_{\mu,\sigma^2}(\sqrt{n}\,|\overline{X}_{(n)} - \mu| \leq \sqrt{n}\ell)\, P_{\mu,\sigma^2}(N = n)$$

$$= \sum_{n=n_1}^{\infty} P_{\mu,\sigma^2}(\sqrt{n_1}\,|\overline{X}_{(n_1)} - \mu| \leq \sqrt{n}\ell) P_{\mu,\sigma^2}(N = n)$$

since every $\sqrt{n}(\overline{X}_{(n)} - \mu)$ is $\mathcal{N}(0, \sigma^2)$ - distributed

$$= P_{\mu,\sigma^2}(\sqrt{n_1}\,|\overline{X}_{(n_1)} - \mu| \leq \sqrt{N}\ell)$$

14

$$\geq P_{\mu,\sigma^2}\left(\sqrt{n_1}\mid \overline{X}_{(n_1)} - \mu\mid\, \leq \ell\sqrt{S^2_{n_1} t^2_{n_1-1;\alpha/2}/\ell^2}\right)$$

according to the choice of N

$$= P_{\mu,\sigma^2}\left(\sqrt{n_1}\mid \overline{X}_{(n_1)} - \mu\mid\,/S_{n_1} \leq t_{n_1-1;\alpha/2}\right) = 1 - \alpha$$

since $\sqrt{n_1}(\overline{X}_{(n_1)} - \mu)/S_{n_1}$ has a central t_{n_1-1} – distribution,

i.e. $[\overline{X}_{(N)} - \ell; \overline{X}_{(N)} + \ell]$ is a sequential confidence interval at level $1 - \alpha$ for μ. $\qquad\square$

All these examples have in common that, on the one hand, after each observation one decides whether the sampling shall be stopped or continued and, on the other hand, one specifies the terminal decision which is taken if sampling is stopped.

The following definition gives a more general and mathematically precise formulation of this kind of procedure (see [Ir]):

(1.12) Definition

a) A <u>sequential statistical experiment</u> $\mathcal{E}$ is a tuple

$$((\mathcal{X}, \mathcal{B}),\ \Theta, (P_\vartheta)_{\vartheta\in\Theta}, T, (\mathcal{B}_t)_{t\in T})$$

consisting of

- *a measurable space $(\mathcal{X}, \mathcal{B})$, the <u>sample space</u>,*
- *a non-empty set Θ, the <u>parameter space</u>,*
- *a family $(P_\vartheta)_{\vartheta\in\Theta}$ of <u>probability distributions</u> on $(\mathcal{X}, \mathcal{B})$; P_ϑ, $\vartheta \in \Theta$, being the probability distributions admitted under the model assumptions,*
- *a non-empty set $T \subset \mathrm{I\!N}_0$ of possible <u>time points for observations</u>, and*
- *an isotonic family $(\mathcal{B}_t)_{t\in T}$ of sub- σ-algebras of $\mathcal{B}$, $\mathcal{B}_t$ representing the <u>information</u> available up to time t.*

b) A <u>sequential statistical decision problem</u> $\mathcal{T}$ is a tuple

$$(\mathcal{E}, (D, \mathcal{D}), L, C)$$

where

- *$\mathcal{E}$ is a sequential statistical experiment,*
- *$(D, \mathcal{D})$ is a measurable space, the <u>decision space</u>; the elements $d \in D$ are <u>terminal decisions</u>,*
- *L is a function $L : \Theta \times D \to [0; \infty)$, the <u>loss function</u>,*
- *C is a function $C : \Theta \times \mathrm{I\!N}_0 \to [0;\infty]$ such that $C(\vartheta, 0) = 0\ \forall \vartheta \in \Theta$, $C(\vartheta, \cdot)$ is monotonically increasing for each $\vartheta \in \Theta$; C is the <u>sampling cost function</u>.*

c) A <u>purely sequential decision procedure</u> for a sequential statistical decision problem $\mathcal{T}$ is a pair

$$(N, \varphi)$$

consisting of

> – a *stopping rule* N *(with respect to* $(\mathcal{B}_t)_{t\in T}$*) i.e. a function* $N : \mathcal{X} \to T \cup \{\infty\}$ *such that* $\{N = t\} \in \mathcal{B}_t \ \forall t \in T$*, and*
>
> – *a family* $\varphi = (\varphi_t)_{t\in T}$ *of randomized terminal decision functions* φ_t *with respect to* $\mathcal{B}_t$*, i.e. stochastic kernels* φ_t *between* $(\mathcal{X}, \mathcal{B}_t)$ *and* $(D, \mathcal{D})$*;* φ *is called* <u>*terminal decision procedure*</u>*.*

For a further explanation of these notions we add some remarks:

(1.13) Remarks

a) The assumption that the observed data are random variables whose distribution is not completely known, but is one of the distributions $P_\vartheta, \vartheta \in \Theta$, given by the model specification, is fundamental for the whole field of statistics.

b) For practical purposes the cases

$$T = \{1, \ldots, n\} \ \text{(observations with } horizon \ n)$$
$$T = \text{IN or IN}_0 \ \text{(potentially unbounded sample sizes)}$$

are of special importance.

c) In most cases the observed data are realizations of a sequence $(X_t)_{t\in T}$ of random variables

$$X_t : (\Omega, \mathcal{S}) \to (\text{IR}^k, I\!\!B^k)$$

i.e. $(\mathcal{X}, \mathcal{B}) = (\text{IR}^{k\cdot T}, I\!\!B^{k\cdot T})$. Since the underlying space $(\Omega, \mathcal{S})$ has no influence on the decision problem – the induced distributions are what matters – one will assume $(\Omega, \mathcal{S}) = (\mathcal{X}, \mathcal{B})$ and consequently $X_t = \pi_t$ (projection on the t-th component); then one obtains in particular

$$\mathcal{B}_t = \sigma(X_s : s \leq t).$$

d) $(\mathcal{B}_t)_{t\in T}$ is called isotonic iff $\mathcal{B}_s \subset \mathcal{B}_t$ for $s \leq t$. The assumption of isotony means that one has, at later time points $t \in T$, at least as much information at one's disposal as at earlier time points $s \in T$. If the $x \in \mathcal{X}$ are obtained (as in part c)) from random variables $X_t : (\Omega, \mathcal{S}) \to (\text{IR}^k, I\!\!B^k)$ one has

$$\mathcal{B}_t = (\pi_{k\cdot T_t})^{-1}(I\!\!B^{k\cdot T_t}) \ \text{where} \ T_t := \{s \in T : s \leq t\},$$

i.e. at time t the first $k\cdot \mid T_t \mid$ components of $x \in \mathcal{X}$ are known.

e) The loss function L and the cost function C are of special importance if the *qualities* of different statistical procedures have to be compared.

f) The decisions whether to stop sampling and which terminal decision has to be taken may depend only on the present but not on the future information. This is just made precise by the $\mathcal{B}_t$-measurability assumptions on N and φ_t.

g) In most applications only sequential decision procedures (N, φ) with the *termination property*

$$P_\vartheta(N \in T) = 1 \quad \forall \vartheta \in \Theta$$

are of practical importance (but see footnote 14).

16

h) For a "sample" $x \in \mathcal{X}$, a purely sequential decision procedure (N, φ) works in the
following way:
(i) Sampling is stopped at time $N(x)$ (if $N(x) \in T$).
(ii) Using $\varphi_{N(x)}(\cdot, x)$ the terminal decision is made.
(iii) This leads to sampling costs $C(\vartheta, N(x))$ and (expected) loss

$$\int L(\vartheta, e)\varphi_{N(x)}(de, x).$$

i) Randomized terminal decision functions where the φ_t are one-point (Dirac-) distributions on terminal decisions $d \in D$ deserve special interest; obviously, these
functions may be identified with *non-randomized* terminal decision functions

$$\delta_t : \mathcal{X} \to D, \ \mathcal{B}_t\text{-measurable}.$$

To illustrate that this rather complicated mathematical model of processing sequentially
may serve to describe quite concrete statistical procedures we fit the examples (1.7) -
(1.10) into this formalism:

(1.14) Examples

According to the (verbal) descriptions of the respective problems we formulate:

a) *Curtailed inspection plan (see (1.7))*

$(\mathcal{X}, \mathcal{B}) = (\{0,1\}^n, \mathcal{P}(\{0,1\}^n)), \ \Theta = [0;1], \ P_\vartheta = \otimes_{i=1}^n \mathcal{B}(1,p) =: (\mathcal{B}(1,p))^n$

$T = \{1, \ldots, n\}, \ \mathcal{B}_t = \pi_{\{1,\ldots,t\}}^{-1}(\mathcal{P}(\{0,1\}^t))$ (see (1.13) d))

$D = \{ \text{accept, reject} \}, \ \mathcal{D} = \mathcal{P}(D).$

Loss and cost functions are not specified in (1.7) (reasonable assumptions could
be

$$L(\vartheta, d) = \begin{cases} K_1 & \vartheta < \vartheta_0 \text{ and } d = \text{reject} \\ K_2 & if \quad \vartheta > \vartheta_0 \text{ and } d = \text{accept} \\ 0 & \text{elsewhere,} \end{cases}$$

where $\vartheta_0 \in (0;1)$ is a fixed critical value, and $C(\vartheta, i) = c \cdot i$, i.e. linear sampling
costs).
Then the curtailed inspection plan $(\widehat{n,a})$ is just the purely sequential decision
procedure defined by the stopping rule

$$N = \min\{m \le n : S_m := \sum_{i=1}^m X_i \in \{a+1, m+a-n\}\}$$

and the non-randomized terminal decision functions (see (1.13) i))

$$\delta_m = \begin{cases} \text{accept} & \le a \\ & \text{if } S_m \\ \text{reject} & \ge a+1 \end{cases} \ , \ 1 \le m \le n.$$

b) *SPRT for the iid case (see (1.8))*

$(\mathcal{X}, \mathcal{B}) = (\mathrm{IR}^{\mathrm{IN}}, \mathcal{B}^{\mathrm{IN}})$, $\Theta = \{1, 2\}$, $P_\vartheta = \otimes_{i=1}^{\infty} Q_\vartheta =: (Q_\vartheta)^{\mathrm{IN}}$ where $Q_{\vartheta_1} \neq Q_{\vartheta_2}$,

$T = \mathrm{IN}$, $\mathcal{B}_t = \pi_{\{1,\dots,t\}}^{-1}(\mathcal{B}^t)$ (see (1.13)d))

$D = \{d_1, d_2\}$ where $d_j \sim$ accept H_j; $\mathcal{D} = \mathcal{P}(D)$.

Again loss and cost functions are not yet specified; here

$$L(\vartheta, d) = \begin{cases} s_j & \text{if } \vartheta = j,\ d = d_{3-j}, j = 1, 2, \\ 0 & \text{elsewhere} \end{cases}$$

and $C(\vartheta, n) = c \cdot n$ seem to be reasonable. Using the notions introduced in (1.8)

$$f_j \text{ for the densities of } Q_j$$

and

$$q, q_n \text{ for the likelihood ratios}$$

the sequential probability ratio test δ_{k_1, k_2} is the purely sequential decision procedure defined by the stopping rule

$$N := \inf\{n \in \mathrm{IN} : q_n \notin (k_1, k_2)\}, \quad (\inf \emptyset := \infty)$$

and the non-randomized terminal decision functions

$$\delta_n = \begin{cases} 1 & \leq 1 \\ & \text{if } q_n \\ 2 & > 1 \end{cases}, \quad n \in \mathrm{IN}.$$

c) *Stein's two-stage procedure (see (1.10))*

$(\mathcal{X}, \mathcal{B}) = (\mathrm{IR}^{\mathrm{IN}}, \mathcal{B}^{\mathrm{IN}})$, $\Theta = \mathrm{IR}^1 \times \mathrm{IR}_+^1$, $P_{(\mu, \sigma^2)} = \otimes_{i=1}^{\infty} \mathcal{N}(\mu, \sigma^2) =: (\mathcal{N}(\mu, \sigma^2))^{\mathrm{IN}}$,

$T = \mathrm{IN}$, $\mathcal{B}_t = \pi_{\{1,\dots,t\}}^{-1}(\mathcal{B}^t)$, $D = \{(\underline{g}, \overline{g}) \in \mathrm{IR}^2 : \overline{g} - \underline{g} = 2\ell\}$, $\mathcal{D} = \mathcal{B}_{|D}^2$

and

$$L((\mu, \sigma^2), (\underline{g}, \overline{g})) = \begin{cases} 1 & \not\ni \\ & \text{if } [\underline{g}, \overline{g}] \quad \mu, \\ 0 & \ni \end{cases}$$

$c((\mu, \sigma^2), n) = c \cdot n$.

For constants $\alpha \in (0; 1)$ and $n_1 \in \mathrm{IN}$, $n_1 \geq 2$, the corresponding two stage procedure of Stein is defined by the stopping rule

$$N := \max\{n_1, \lceil S_{n_1}^2 \cdot t_{n_1 - 1; \alpha/2}^2 / \ell^2 \rceil\}$$

and the non-randomized terminal decision function

$$\delta_n(x_1, \dots, x_n) := (\overline{x}_{(n)} - \ell,\ \overline{x}_{(n)} + \ell). \qquad \square$$

The mathematical model of (1.12) therefore turns out to be flexible enough to cover very different sequential procedures and to make this (verbal) description precise.

18

§ 5 Objections to purely sequential statistical decision procedures

Although proceeding sequentially seems to be very reasonable for statistical investigations and though the definitions of (1.12) – in particular splitting up sequential procedures into a stopping rule and a terminal decision function – turn out to be useful for many situations, some aspects seem not to be taken into account which are of importance for practical applications of sequential analysis. Describing the sequential component by a stopping rule one can only register the *end* of sampling but not the *kind* of sampling; e.g. it is not determined whether sampling takes place one-at-a-time or in groups. The reason is that a stopping rule gives, at any time point, only the advice whether to stop or not but not *how* the next observations have to be made which is a *design* aspect of the experiment. This is of importance if the sampling cost function $C(\vartheta, t)$ is non-linear in t. To illustrate this aspect we again consider example (1.1)/(1.7):

(1.15) Remark/Example
For most routine investigations (e.g. quality control, pharmaceutical serial experiments) preparing, carrying out and evaluating k single experiments turn out to be much more time- and money-consuming than to explore a sample of size k.

If one has e.g. in (1.7) besides the inspection costs c per item additional fixed costs c_0 for preparing and evaluating a (sub-) sample (comp. e.g. [Ab 1], [Ma]) k successive one-at-a-time observations lead to sampling costs of $k(c_0 + c)$ whereas a (combined) sample of size k causes sampling costs $c_0 + kc$. Due to this effect the saving of sample numbers, yielded by proceeding sequentially, may be more than compensated by additional costs.

If the cost function in (1.7)/(1.14)a) has the form

$$C(\vartheta, i) = c_0 + c \cdot i \quad \text{where } c_0 = c$$

one obtains for the expected *costs* of the sampling plans $(20, 2)$ and $(\widehat{20, 2})$ the values given in figure 1.7:

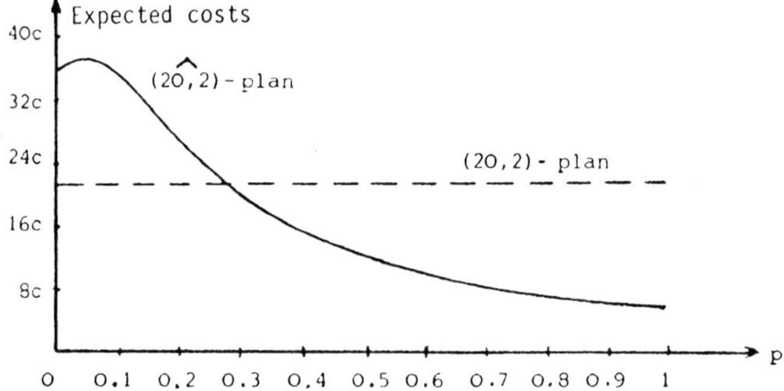

Figure 1.7: Expected sampling costs of $(20, 2)$ and $(\widehat{20, 2})$
for the cost function $c \cdot (i + 1)$

In particular for "small" values of p, which (hopefully) are important for practical quality control, the curtailed inspection plan $(\widehat{20, 2})$ turns out to be considerably more expensive than the classical $(20, 2)$ plan – hence the expected sample size (ASN) is not the appropriate scale for comparing inspection plans. $\qquad\square$

Closely related to this objection to purely sequential procedures is the following aspect:

(1.16) Remark
In many agricultural investigations a single observation needs a full vegetation period. Obviously it makes no sense to observe only *one* plant per year and that (so many years!) until the gathered data are informative enough. In addition to high fixed costs, this kind of experimentation would lead to an undue loss of time. $\qquad\square$

Describing the sequential component of a decision procedure by a stopping rule has, therefore, no regard to the – often advantageous – possibilities of choosing, during the course of investigation, the size of the next sub-sample or of parallel processing of observations. This might be one of the (implicit) reasons that sequential procedures are used rather seldom in practical applications.

A further reservation against purely sequential decision procedures could be caused by the (vague) fear that very small sample sizes – which can occur with sequential procedures – are the result of "outliers" (but not of real differences in the treatments).

§ 6 Sequentially planned statistical procedures

The foregoing arguments/objections lead to the question whether it is possible to retain, by suitable modifications, the advantages of purely sequential procedures without taking into the bargain the disadvantages.

To motivate the concept, developed in the sequel, we again consider the example $(1.1)/(1.7)/(1.15)$ from quality control:

(1.17) Example
Let for an acceptance-rejection quality control (see $(1.1)/(1.7)$) the inspection costs be given by
$$c_0 + c \cdot i, \quad c_0, c > 0$$
(c_0 fixed costs, c inspection costs per item). As figure 1.8 shows, then the following dilemma arises:

(i) On the one hand one should use the information gathered during the course of inspection to save – in particular for "large" values of p – costs by early decisions.

(ii) On the other hand one-at-a-time sampling, which would allow for terminal decisions after each single inspection, leads to unreasonable high costs (due to the accumulation of fixed costs) for "small" values of p.

Furthermore, a terminal decision cannot be reached earlier than 3 observations (and that iff the first 3 items are defective). Therefore, it seems reasonable to start with an *initial sample* of at least size 3 (compared with one-at-a-time sampling this saves some fixed costs) and then allow the future course depend on the result of these inspections. This idea to investigate *groups* of items whose sizes depend on the present information may then be used again.

Consider, for instance, the following (ad hoc) 3-stage procedure G

(1.18)

1. Take an initial sample of size 10 and inspect the items; if
   (i) $S_{10} \geq 3$ reject the batch
   (ii) $S_{10} = 2$ ⎫
   (iii) $S_{10} = 1$ ⎬ take a second sample of size ⎧ 5
   (iv) $S_{10} = 0$ ⎭   ⎨ 10
                                                      ⎩ 8

2. If in case
   1.(ii) $S_{15} \geq 3$ reject the batch, elsewhere take a third sample of size 5

   1.(iv) $S_{18} \geq 3$ reject the batch,

   $S_{18} = 0$ accept the batch,

   elsewhere take a third sample of size 2

3. If $S_{20} \leq 2$ accept the batch and if $S_{20} \geq 3$ reject it.

G leads to the expected sampling costs given in figure 1.8.

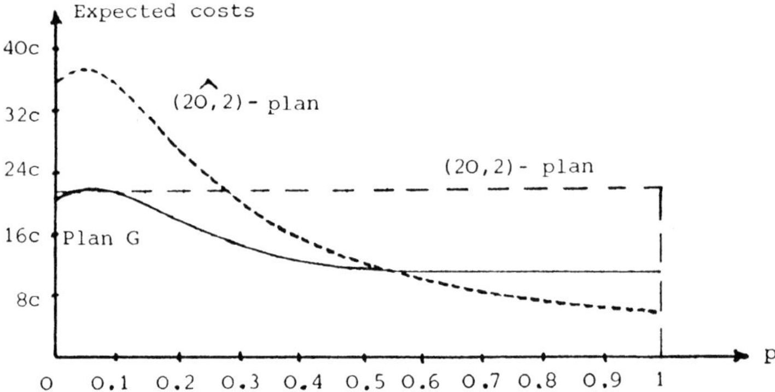

Figure 1.8: Expected sampling costs for the
inspection plans $(20,2), (\widehat{20,2})$ and G.

The plan G has the same rejection probabilities as the classical $(20,2)$ plan; it avoids the disadvantages of curtailed inspection arising for small values of p, but nevertheless allows an early detection of large values of p (i.e. poor production quality).   □

In a similar way it seems, for the biometrical investigations of example (1.16), reasonable to choose the (initial) sample size large enough to guarantee under normal circumstances the desired statistical accuracies (level requirements, restrictions on the width of confidence intervals...) and make additional experiments only if some data are missing (due to external influences).

These remarks suggest to investigate statistical procedures which on the one hand allow, by proceeding sequentially, early decisions in case of very informative data but, on the other hand, avoid, by group sequential and parallel observations, the disadvantages of purely sequential procedures. Since here also the size of the next (sub-) sample will be determined / planned according to the information gathered up to that time we will call these procedures *sequentially planned.*

(1.19) Remarks

Wald was aware that proceeding purely sequentially could cause practical difficulties and that therefore group-sequential procedures might be better suited ("For practical reasons it may sometimes be preferable to take the observations in groups, rather than singly" [Wa], p. 101). But he considered only "fixed" group sizes and investigated only the case of equal group sizes in some detail ([Wa], pp. 102-104; see also [Gh], pp. 224-228). Stein's two-stage procedure (see (1.10)), however, lets already foreshadow the basic idea of sequentially planned procedures (determination of the size of (sub-) samples according to the available information). In the meantime several other authors pointed out that sequential procedures with variable group sizes could be useful (see e.g. [Ab 3], [Be] p. 355, [E], [Hay], [S/E], [Wal 1]). A systematic investigation of such methods and their use in form of practicable procedures seem to be missing; an attempt to develop such procedures will be made in the sequel. □

There are several possibilities to define sequentially planned procedures (see appendix A). Our model will be based on the concept of "control variables" due to Haggstrom [Ha] which was originally developed for problems of sequential design of experiments; moreover a close relation to definition (1.12) will be established.

(1.20) Definition

a) *For* $M \subset \mathbb{N}$ *let denote*[17]

$$
\begin{aligned}
A_M \ &:= \ \cup_{j=1}^{\infty} M^j \cup \{()\} \\
&= \ \{(a_1, \ldots, a_j) : j \in \mathbb{N}; a_i \in M \ \forall 1 \leq i \leq j\} \cup \{()\} \\
A_M^{\star} \ &:= \ M^{\mathbb{N}} = \{(a_1, a_2, \ldots) : a_i \in M \ \forall i \in \mathbb{N}\}.
\end{aligned}
$$

b) *If* $a = (a_1, \ldots, a_j)$, $b = (b_1, \ldots, b_k) \in A_M$ *then* $a \prec b$ *iff* $j \leq k$ *and* $a_i = b_i$
$\forall 1 \leq i \leq j$*; if* $a = (a_1, \ldots, a_j) \in A_M$, $b = (b_1, b_2, \ldots) \in A_M^{\star}$ *then* $a \preceq b$ *iff* $a_i = b_i$
$\forall 1 \leq i \leq j$*; if* $a = (a_1, a_2, \ldots), b = (b_1, b_2, \ldots) \in A_M^{\star}$ *then* $a \preceq b$ *iff* $a_i = b_i \ \forall i \in \mathbb{N}$
(in particular $() \preceq a$ *holds* $\forall a \in A_M \cup A_M^{\star}$*).*

For a better understanding of these notions we mention:

(1.21) Remarks

a) The set M is used to describe the sizes of (sub-) samples which can be drawn in the respective decision problem[18]; important special cases are e.g.

---

[17]The index M will be omitted when there is no possibility of misunderstanding.

[18]To avoid further terminological complications we assume the same set M for all stages; (external) restrictions will be regarded when defining sampling plans.

$M = \{1, \ldots, m\}$: sub-samples whose sizes do not exceed m

$M = \mathbb{IN}$: no a priori restrictions on the size of sub-samples

$M = \{2n : 1 \leq n \leq m\}$: only pairwise observations are possible; not more than m pairs per sub-sample are allowed.

b) The set A contains all possible finite sequences of sizes of (sub-) samples: $a = (a_1, \ldots, a_j) \in A$ means that one starts with an initial sample of size a_1, continues with a further sample of size a_2 etc. up to a j-th sample of size a_j; () means that no observation is made. Obviously the set A is countable. We will call the $a \in A \cup A^*$ *sample sequences*.

c) For $a = (a_1, \ldots, a_j) \in A$

$$h(a) := j \quad (\text{with } h(()) = 0)$$

gives the *number* of sub-samples (*number of stages*) and

$$g(a) := \sum_{i=1}^{j} a_i \quad (\text{with } g(()) = 0)$$

gives the total number of observations (*total sample size*). Therefore

$$A_k := \{a \in A : h(a) = k\}$$

describes the set of all k-stage sample sequences whereas

$$A^k := \{a \in A : g(a) = k\}$$

contains all sample sequences with total sample size k.

d) The relation $\preceq$ is a partial order on $A \cup A^*$; $a \preceq b$ means that the sequence b is a continuation of a.

e) $b \in A$ is called *direct successor* of $a \in A$ if $a \prec b$ but no $c \in A$ exists such that $a \prec c \prec b$; for $a = (a_1, \ldots, a_j) \in A$ just the sequences

$$ak := (a_1, \ldots, a_j, k) \quad \text{where } k \in M$$

are the direct successors of a. For $a_1, a_2 \in A$ with $a_1 \npreceq a_2 \npreceq a_1$ one obtains also for all $b_1, b_2 \in A \cup A^*$ with $a_1 \preceq b_1, a_2 \preceq b_2$ that $b_1 \npreceq b_2 \npreceq b_1$, i.e. $(A, \preceq)$ is a *tree* (see figure 1.9):

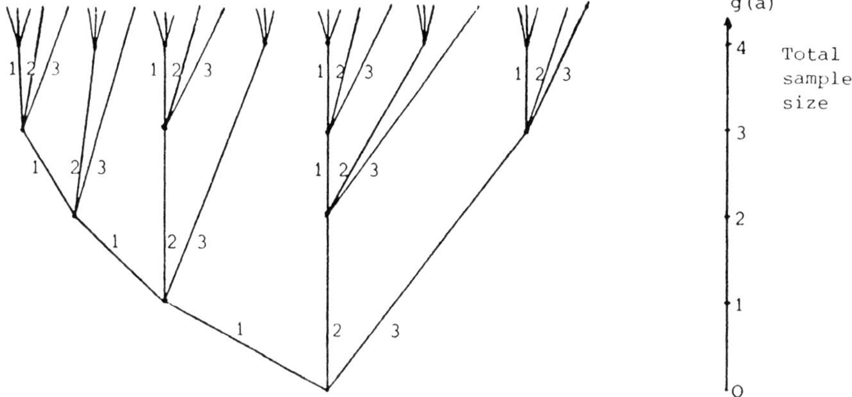

Figure 1.9: First part of the tree $(A, \preceq)$ for $M = \{1, 2, 3\}$

Statistical decision procedures where the sizes of consecutive sub-samples are also determined sequentially may now be described in the following way:

(1.22) Definition

a) A general sequential statistical decision problem $\mathcal{P}$ is a tuple

$$((\mathcal{X}, \mathcal{B}), \Theta, (P_\vartheta)_{\vartheta \in \Theta}, \mathcal{A}, (\mathcal{B}_a)_{a \in A}, (D, \mathcal{D}), (L_a)_{a \in \mathcal{A}}, c)$$

consisting of

- *a measurable space $(\mathcal{X}, \mathcal{B})$,*

- *a non-empty set Θ, the parameter space,*

- *a family $(P_\vartheta)_{\vartheta \in \Theta}$ of probability distributions on $(\mathcal{X}, \mathcal{B})$*

- *a set $\mathcal{A} \subset A = \cup_{j=1}^\infty M^j \cup \{()\}$ (where $M \subset \mathbb{N}$) of admissible finite sample sequences,*

- *an "isotonic" family $(\mathcal{B}_a)_{a \in A}$ of sub-σ-algebras of $\mathcal{B}$, i.e. $\mathcal{B}_a \subset \mathcal{B}_b \subset \mathcal{B} \; \forall a, b \in A : a \prec b$,*

- *a measurable space $(D, \mathcal{D})$, the decision space,*

- *a family $(L_a)_{a \in \mathcal{A}}$ of functions*

$$L_a : \Theta \times D \times \mathcal{X} \to [0; \infty]$$

which for each $\vartheta \in \Theta$ and $d \in D$ are $\mathcal{B}_a$-measurable, the family of loss functions,

- *a function $c : \Theta \times M \to [0; \infty)$ such that $c(\vartheta, \cdot)$ is monotonically increasing for each $\vartheta \in \Theta$, the sampling cost function.*

b) A sequentially planned statistical decision procedure (for a general sequential statistical decision problem $\mathcal{P}$) is a pair

$$(\tau, \varphi)$$

consisting of

- *a function $\tau : \mathcal{X} \to A \cup A^*$ such that $\{\tau \preceq a\} \in \mathcal{B}_a$, $\{\tau \succeq (a_1, \ldots, a_j, k)\} \in \mathcal{B}_a$ for all $a = (a_1, \ldots, a_j) \in A$ and all $k \in M$; τ is called a (non-randomized) <u>sequential sampling plan</u>, and*

- *a family $\varphi = (\varphi_a)_{a \in A}$ of randomized terminal decision functions φ_a with respect to $\mathcal{B}_a$; φ is called <u>terminal decision procedure</u>.*

The special case $M = \{1\}$ – i.e. after each single observation a new decision is made – leads (by the identification of $\underbrace{(1, \ldots, 1)}_{n\text{-times}}$ and n) just back to the notions of (1.12). Therefore most terms of definition (1.22) need no further explanation. From the intuitive meaning of the set A (see (1.21)) we obtain the

(1.23) Remarks

a) Quite similar to the purely sequential case the σ-algebra $\mathcal{B}_a$ represents the information available due to the sample sequence a.

b) The $\mathcal{B}_a$-measurability assumptions on sequential sampling plans again have the meaning that the decision whether to stop with the sample sequence or what size the next (sub-) sample should have may depend only on the information gathered up to that stage. An analogous interpretation holds true for the $\mathcal{B}_a$-measurability of the terminal decision functions φ_a.

c) By the set $\mathcal{A}$ restrictions are taken into account which are given by external conditions on the sample sequences.

$$\mathcal{A} = \{a \in A : h(a) \leq k\}$$

means e.g. that only sample sequences with at most k stages are allowed; similarly

$$\mathcal{A} = \{a \in A : g(a) \leq k\}$$

says that the total number of observations must not exceed k. According to the meaning of $\mathcal{A}$ we extend the family of loss functions by

$$L_a \equiv \infty \quad \forall a \notin \mathcal{A}.$$

For practical purposes only sequential sampling plans τ with the property

$$P_\vartheta(\tau \in \mathcal{A}) = 1 \quad \forall \vartheta \in \Theta$$

are of importance, i.e. sampling plans which a.s. stop sampling with an admissible finite sample sequence.

d) $c(\vartheta, n)$ means the costs for a (sub-) sample of size n if P_ϑ is the underlying probability distribution. A sample sequence $a = (a_1, \ldots, a_j) \in A$ therefore leads to total costs of sampling

$$C(\vartheta, a) := \sum_{i=1}^{j} c(\vartheta, a_i) \ (\text{where } C(\vartheta, ()) = 0)$$

if ϑ is the "true" parameter value.

It is quite easy to *apply* a sequentially planned statistical decision procedure: First an initial sample of size a_1, prescribed by τ, is drawn (since in general no information is available before starting the experiment, i.e. $\mathcal{B}_{()} = \{\emptyset, \mathcal{X}\}$ holds, a_1 will be a *deterministic* value). On the basis of the observed data, which are one- or more-dimensional measurements $x_1, \ldots, x_{a_1}$, one then decides according to the sampling plan τ

- whether sampling is stopped, or

- of which size a_2 the second (sub-) sample will be; a_2 is i.g. a random variable because it is determined in a $\mathcal{B}_{a_1}$-measurable way (namely as a function of $x_1, \ldots, x_{a_1}$).

In the first case the terminal decision given by φ_{a_1} is made. In the second case a further sample of size a_2 is drawn and on the basis of the gathered total information – i.g. the observational data $x_1, \ldots, x_{a_1}, x_{a_1+1}, \ldots, x_{a_1+a_2}$ – again the decision (prescribed by τ) is made whether to stop or how to continue sampling. $\qquad\qquad\Box$

For an illustration we give some examples.

(1.24) Examples

a) The purely sequential decision procedures (N, φ) defined in (1.12) are (by the identification of $\underbrace{(1, \ldots, 1)}_{n\text{-times}}$ and n) the special case of sequentially planned procedures specified by $M = \{1\}$ and $\mathcal{A} = A$. In particular the procedures mentioned in (1.14) (curtailed inspection plan, SPRT for iid random variables, Stein's two-stage procedure) turn out to be special examples of such sequentially planned procedures.

b) But Stein's two-stage procedure also in another way fits into this model: The initial sample has fixed size n_1 (therefore only once a sample has to be drawn, hence e.g. a parallel investigation of the items is possible), and also the second sample – with a random size – can be drawn in one step. Using

$$(\mathcal{X}, \mathcal{B}), \Theta, (P_\vartheta)_{\vartheta \in \Theta}, (D, \mathcal{D}), \alpha, n_1$$

with the same meaning as in (1.14)c), $Y := \lceil S_{n_1}^2 \cdot t_{n_1-1, \alpha/2}^2 / \ell^2 \rceil$ and $M = \mathrm{I\!N}$, Stein's two-stage procedure may be described as sequentially planned statistical decision procedure by

$$\tau = \begin{cases} (n_1) & \leq n_1 \\ & \text{if } Y \\ (n_1, Y - n_1) & > n_1 \end{cases}$$

26

and

$$\varphi_a(x) = (\overline{x}_{(g(a))} - \ell, \ \overline{x}_{(g(a))} + \ell).$$

This is of special importance for non-linear sampling costs – e.g. costs of the form $c_0 + c \cdot i$ where $c_0 > 0$ means fixed costs (comp. (1.15)); in this case one has to pay only

$$c_0 + cn_1 + \left(c_0 + c \cdot (Y - n_1)\right) 1_{\{Y > n_1\}}$$

instead of the (much) higher costs

$$(c_0 + c)n_1 + (c_0 + c)(Y - n_1) \, 1_{\{Y > n_1\}}$$

occuring at one-at-a-time sampling.

The remarks made under point b) possibly give an explanation why practitioners prefer two-stage procedures to purely sequential ones.

c) The three-stage procedure of Hall [Hal] tries to combine the advantages of Stein's two-stage procedure with those of purely sequential procedures (in particular with respect to the asymptotic efficiency). In the situation of example b) let[19] $Y_1 := \lfloor c \, S_{n_1}^2 \cdot u_{\alpha/2}^2 / \ell^2 \rfloor + 1$, where $c \in (0, 1)$ is a constant,

$$\tau = \begin{cases} (n_1) & \text{if } Y_1 \leq n_1 \\ (n_1, Y_1 - n_1) & \text{if } n_1 < Y_1 \geq Y_2 := \lfloor S_{Y_1}^2 \cdot u_{\alpha/2}^2 / \ell^2 \rfloor + 1 \\ (n_1, Y_1 - n_1, Y_2 - Y_1) & \text{if } n_1 < Y_1 < Y_2, \end{cases}$$

and $\varphi_a(x) = (\overline{x}_{(g(a))} - \ell, \overline{x}_{(g(a))} + \ell)$. This sequentially planned procedure ("triple sampling for sequential estimation of a mean") behaves asymptotically nearly as "well" as the purely sequential procedure of Chow/Robbins [C/R] (see [Hal]) and has considerable advantages concerning sampling costs if fixed costs occur.

d) The "double sampling inspection procedure" of Dodge and Romig ([D/R], p. 23)

"(a) Inspect a first sample of n_1 pieces.

(b) If the number of defectives found in the first sample does not exceed c_1, the acceptance number for the first sample, accept the lot.

(c) If the number of defectives found in the first sample exceeds c_2, the acceptance number for the combined first and second samples, inspect all the pieces in the remainder of the lot.

(d) If the number of defectives found in the first sample exceeds c_1 but does not exceed c_2, inspect a second sample of n_2 pieces.

(e) If the total number of defectives found in the first and second samples combined does not exceed c_2, accept the lot.

(f) If the total number of defectives found in the first and second samples combined exceeds c_2, inspect all the pieces in the remainder of the lot.

[(g) Correct or replace all defective pieces found.]"

---

[19] $\lfloor x \rfloor$ describes the largest integer $\leq$ x.

also fits into the general model (1.22): Let $(\mathcal{X}, \mathcal{B}), \Theta, (P_\vartheta)_{\vartheta \in \Theta}$ be as in (1.14)a),

$$D = \{\text{ acception, total inspection }\}, \mathcal{D} = \mathcal{P}(D),$$
$$M = \{1, \ldots, n\} \ (n \sim \text{size of the batch})$$

and $n_1, n_2, c_1, c_2 \in \mathrm{I\!N}, c_1 < c_2$, be constants. Then the sequential sampling plan

$$\tau = \begin{cases} (n_1) & \text{if } \sum_{i=1}^{n_1} X_i \leq c_1 \\ (n_1, n - n_1) & \text{if } \sum_{i=1}^{n_1} X_i > c_2 \\ (n_1, n_2) & \text{if } c_1 < \sum_{i=1}^{n_1} X_i \leq c_2 \end{cases}$$

and the (non-randomized) decision procedure

$$\varphi_a(x_1, \ldots, x_{g(a)}) = \begin{cases} \text{acceptance} & \leq c_2 \\ & \text{if } \sum_{i=1}^{g(a)} x_i \\ \text{total inspection} & > c_2 \end{cases}$$

just describe the double sampling inspection procedure. $\qquad\qquad \square$

In particular, for reductions by sufficiency, transitivity, and invariance *randomized* sampling plans are needed:

(1.25) Definition
Let
$$\mathcal{P} = ((\mathcal{X}, \mathcal{B}), \Theta, (P_\vartheta)_{\vartheta \in \Theta}, \mathcal{A}, (\mathcal{B}_a)_{a \in A}, (D, \mathcal{D}), (L_a)_{a \in A}, c)$$

be a general sequential decision problem.

a) A randomized sequential sampling plan $\beta = (\beta_a)_{a \in A}$ is a family of transition probabilities β_a from $(\mathcal{X}, \mathcal{B}_a)$ to $(M_0, \mathcal{P}(M_0))$ where[20] $M_0 := M \cup \{0\}$:

$$\beta_a : \mathcal{X} \times \mathcal{P}(M_0) \to [0; 1].$$

(For simpler notations let $\beta_a(x, k) := \beta_a(x, \{k\})$).

b) For a randomized sequential sampling plan $\beta = (\beta_a)_{a \in A}$ and $a = (a_1, \ldots, a_n) \in A$, $n \in \mathrm{I\!N}$, let

$$b_a^\beta(x, k) := b_a(x, k) = \beta_a(x, k) \prod_{i=0}^{n-1} \beta_{(a_1, \ldots, a_i)}(x, a_{i+1})$$

$\forall x \in \mathcal{X}, k \in M_0$, *and[21]* $b_{()} := \beta_{()}$.

c) A randomized sequential sampling plan $\beta = (\beta_a)_{a \in A}$ has the termination property if

$$\sum_{a \in A} b_a^\beta(\cdot, 0) = 1 \quad P_\vartheta - a.s. \ \forall \vartheta \in \Theta.$$

d) A randomized sequentially planned statistical decision procedure is a pair (β, φ) consisting of a randomized sequential sampling plan β and a terminal decision procedure φ.

---

[20]"0" corresponds to the decision "no further observation".
[21]$(a_1, \ldots, a_0) := ()$.

(1.26) Remarks

a) Due to the explanation of definition (1.22) the interpretation of (1.25) is obvious: $\beta_a(x,\cdot)$ gives, for $x \in \mathcal{X}$ and $a = (a_1,\ldots,a_n)$, the "conditional probabilities" of the possible continuations of the experiment – these are represented by M_0 –, if the previous sample sizes were $a_1,\ldots,a_n$. The $\mathcal{B}_a$-measurability of $\beta_a(\cdot,k)$ guarantees that only the information available at $a = (a_1,\ldots,a_n)$ is used for determining the probability distribution on M_0.
The terms $b_a(x,k)$ defined in (1.25)b) describe, for $x \in \mathcal{X}, a \in A$ and $k \in M$, the probability that the realized sample sequence belongs to the sub-tree $\{e \in A : e \succeq ak\}, b_a^\beta(\cdot,0)$ determines the probability that under sampling according to β the sample sequence $a \in A$ will be realized.

b) The property mentioned in (1.25)c) is an analog of (1.13)g) – it means that sampling is stopped after finitely many stages with an admissible sample sequence.

c) Definition (1.25)a) generalizes the previous definition (1.22): Let $\tau : \mathcal{X} \to A \cup A^\star$ be a (non-randomized) sampling plan. Then τ may be identified with the (randomized) sampling plan β given by

$$\beta_a(x,\cdot) := \begin{cases} \varepsilon_j & \text{if there exists a } j \in M \, s.t. \ \tau(x) \succeq aj \\ \varepsilon_0 & \text{elsewhere} \end{cases}$$

(where ε_y denotes the one-point measure in y).

$$\{\beta_a(\cdot,k) = \alpha\} = \left\{ \begin{array}{lll} \emptyset & & \alpha \in (0;1) \\ \{\tau \succeq ak\} & \text{if} & \alpha = 1 \\ \{\tau \succeq ak\}^c & & \alpha = 0 \end{array} \right\} \in \mathcal{B}_a$$

shows that $\beta = (\beta_a)_{a \in A}$ is a well-defined randomized sequential sampling plan. Moreover, obviously

$$b_a^\beta(\cdot,0) = 1_{\{\tau = a\}} \quad \forall a \in A;$$

hence

$$P_\vartheta(\tau \in \mathcal{A}) = 1 \Leftrightarrow \sum_{a \in \mathcal{A}} b_a^\beta(\cdot,0) = 1 \quad P_\vartheta - a.s.,$$

i.e. the notions of termination property coincide. $\qquad\qquad\square$

The aim of the following chapters is to construct "reasonable" – and, indeed, "optimal" whenever possible – sequentially planned decision procedures. To achieve this goal, it will of course be necessary to introduce certain criteria for measuring the "quality" of such procedures.

II. Optimal sequential sampling plans

§ 1   Problems of optimal sampling

Statistical decision procedures where the sizes of sub-samples are determined sequentially (*sequentially planned statistical decision procedures*) are described in definition (1.22) by a sequential plan τ and a terminal decision procedure φ. I.e. similarly to purely sequential decision procedures (see definition (1.12)) such a procedure is split up into

- a part τ which is focused upon the sequential determinations of sub-samples and

- a part $\varphi = (\varphi_a)_{a \in A}$ which exclusively concerns the terminal decisions.

For purely sequential procedures the corresponding partition turned out to be very useful (see e.g. [Ir]): In many cases – e.g. Bayesian decision problems – "optimal" terminal decisions $\varphi_a^\star$ can be determined independently from the stopping rule. Looking for an optimal decision procedure one may therefore restrict attention to compute, for fixed $\varphi^\star$, an optimal stopping rule. For this problem the *theory of optimal stopping* yields a great variety of results (we refer to the monograph "Great Expectations: The Theory of Optimal Stopping" by Y. Chow, H. Robbins and D. Siegmund [C/R/S]).

An obvious idea is, therefore, to develop a corresponding *theory of optimal sampling.* Compared with the general sequential decision problems as formulated in (1.22) the situation is much easier since the underlying probability distribution is assumed to be known - hence the decision space does not occur and the "pay-off"-function simplifies:

(2.1) Definition
  A problem of optimal sampling is a tupel

$$((\mathcal{X}, \mathcal{B}, P), \mathcal{A}, (\mathcal{B}_a)_{a \in A}, (Z_a)_{a \in A})$$

  consisting of

- *a probability space* $(\mathcal{X}, \mathcal{B}, P)$

- *a set* $\mathcal{A} \subset A = \cup_{j=1}^{\infty} M^j \cup \{()\}$ *(where* $M \subset \mathbb{N}$*) of admissible finite sample sequences*

- *an isotonic family* $(\mathcal{B}_a)_{a \in A}$ *of sub-σ- algebras of* $\mathcal{B}$, *and*

- *a family* $(Z_a)_{a \in A}$ *of* $(\mathcal{B}_a, \mathbb{B})$*- measurable random variables* Z_a.

Taking into account the comments on definition (1.22) the interpretation of these notions is rather obvious:

(2.2) Remarks

  a) $(\mathcal{X}, \mathcal{B}, P)$ describes the underlying random effects; the probability measure P is assumed to be known.

b) The meaning of the σ-algebras $\mathcal{B}_a$ was explained in (1.22) already. Z_a is called *pay-off* associated to the sample sequence $a \in \mathcal{A}$; for $a \in (A \cup A^*) \backslash \mathcal{A}$ we define
$$Z_a := -\infty.$$

c) In this chapter we consider – if not explicitly mentioned – non-randomized sampling plans only. For the general theory this is no substantial restriction since it is possible to describe each randomized sampling plan, by extending the sample space $\mathcal{X}$ and the σ-algebras $\mathcal{B}_a$ in an appropriate way, as a non-randomized plan – the extended measurable spaces $(\overline{\mathcal{X}}, \overline{\mathcal{B}_a})$ contain also the information about the outcome of the additional randomizations (see e.g. [Si], [Lü]). $\qquad\qquad\square$

For each sequential sampling plan τ

$$Z_\tau := \left\{ \begin{array}{ll} Z_a & \text{on } \{\tau = a\},\ a \in \mathcal{A} \\ -\infty & \text{elsewhere} \end{array} \right.$$

gives the appertaining pay-off (resp. the corresponding sampling costs). The set A is countable; Z_τ is therefore $(\mathcal{B}, \overline{\mathrm{I\!B}})$ measurable.

Problems of optimal stopping are special problems of optimal sampling – namely the case $M = \{1\}$. One therefore immediately obtains a great variety of examples. But for these problems the problem of *planning* does not occur. To illustrate this aspect we consider a simple example:

(2.3) Example (modified S_n/n-problem)

Let $X_1, X_2, \ldots$ be independent identically distributed random variables s.t.

$$P(X_1 = 1) = 1/2 = P(X_1 = -1);$$

the X_i may be interpreted as fair games. At any decision-time one may determine the number of games one wants to play. If one stops playing this "game" after j continuations, leading to k single observations X_i, the result is valued by the average pay-off per decision, i.e. by

$$\sum_{i=1}^{k} X_i/j.$$

We consider three variants of this "game":

(i) There are no restrictions on the number of decisions and on the number of single observations at each stage.

(ii) One may choose at most m single observations at each stage, but there is no restriction on the number of continuation decisions.

(iii) The total number of observations X_i is restricted by $m \in \mathrm{I\!N}$.

These "games" can be formulated in an obvious way as problems of optimal sampling: Define

- $(\mathcal{X}, \mathcal{B}, P) = (\{-1,1\}^{\mathbb{N}}, \mathcal{P}(\{-1,1\}^{\mathbb{N}}), Q^{\mathbb{N}})$ where

$$Q(\{-1\}) = 1/2 = Q(\{1\}),$$

and $X_i := \pi_i$ (projection onto the i-th component; $x_i = \pi_i(x)$ describes, for $x = (x_1, x_2, \ldots) \in \mathcal{X}$, the result of the i-th observation)

- in case (i): $M = \mathbb{N}$ and $\mathcal{A} = A(= \cup_{j=1}^{\infty} M^j \cup \{()\})$
  in case (ii): $M = \{1, \ldots, m\}$ and $\mathcal{A} = A$
  in case (iii): $M = \{1, \ldots, m\}$ and $\mathcal{A} = \{a \in A : g(a) \leq m\}$,

- $\mathcal{B}_a = \sigma(X_1, \ldots, X_{g(a)}), a \in A$; for a finite sample sequence a the σ-algebra $\mathcal{B}_a$ contains the information on the results of the $g(a)$ single observations, and

- $Z_a = \sum\limits_{i=1}^{g(a)} X_i / h(a)$ for $a \in \mathcal{A}, a \neq ()$, and $Z_{()} = 0$.

Then

$$((\mathcal{X}, \mathcal{B}, P), \mathcal{A}, (\mathcal{B}_a)_{a \in A}, (Z_a)_{a \in \mathcal{A}})$$

describes the situation mentioned above; in case (ii) one obtains for the special case $m = 1$ nothing else than the classical S_n/n- problem. $\qquad\qquad\square$

Analogously to the theory of optimal stopping one will now try to receive an "optimal" (expected) pay-off.

(2.4) Definition

Let $((\mathcal{X}, \mathcal{B}, P), \mathcal{A}, (\mathcal{B}_a)_{a \in A}, (Z_a)_{a \in \mathcal{A}})$ be a problem of optimal sampling.

a) $\hat{T} := \{\tau : \tau$ *sequential sampling plan such that EZ_τ exists$\}$*

is the set of $\underline{admissible}$[22] sequential sampling plans.

b) $v := \sup\limits_{\tau \in \hat{T}} E\, Z_\tau$ *is called the $\underline{value}$ (of the problem).*

c) An admissible sequential sampling plan τ^* such that $E\, Z_{\tau^*} = v$ is called $\underline{optimal}$.

In correspondence with our objectives obviously the following questions arise:

How can the value of a problem of optimal sampling be computed?

Does an optimal sequential sampling plan exist?

How can an optimal sequential sampling plan be computed (in an explicit way)?

Is it possible to approximate problems of optimal sequential sampling by "finite" problems which are accessible to numerical algorithms?

---

[22]There is no relation to the admissibility concept of decision theory; existence only of the expected value in the wide sense is assumed, i.e. $EZ^+ < \infty$ or $EZ^- < \infty$.

32

Treating these questions the following notions turn out to be useful:

$$\begin{aligned}
T &:= \{\tau \in \hat{T} : E\, Z_\tau^- < \infty\}, \\
T_a &:= \{\tau \in T : \tau \succeq a\}, \ a \in A \\
T_{a+} &:= \{\tau \in T : \tau \succ a\}, \ a \in A.
\end{aligned}$$

Maximizing the expected pay-off one obviously may restrict attention to sampling plans $\tau \in T$; for $\tau \in T$ holds

$$P(\{\tau \in \mathcal{A}\}) = 1.$$

T_a consists of all plans $\tau \in T$ with fixed initial part a; T_{a+} consists of those $\tau \in T_a$ which continue at a.

§ 2 Optimal sampling plans for finite horizon

If the total sample size is bounded by, say, $m \in \text{IN}$ (*"finite horizon"*) – i.e. $A \subset \{a \in A : g(a) \le m\}$ – the set A is finite. Then it seems reasonable to compute, according to the basic idea of dynamic programming, an optimal sampling plan by backward induction. To this end let

$$\tilde{A} := \{\tilde{a} \in A : \exists a \in A : \tilde{a} \preceq a\}$$

be the minimal sub-tree of A with () as smallest element which contains A; obviously $\tilde{A}$ is finite. Moreover, we define for $a \in \tilde{A}$

$$\tilde{M}_a := \{j \in M : aj \in \tilde{A}\}$$

and for $a = (a_1, \ldots, a_k) \in A$ and $1 \le i \le k$

$$a^i := (a_1, \ldots, a_i)$$

(first i sub-samples).

Using these notations we are now able to answer the questions mentioned above:

(2.5) Theorem
Consider a problem of optimal sampling with finite A. Define for
$a = (a_1, \ldots, a_k) \in \tilde{A}$

$$\begin{aligned}
U_a &:= Z_a \ \text{if}\ \tilde{M}_a = \emptyset, \\
U_a &:= \max\{Z_a, \max_{j \in \tilde{M}_a} E(U_{aj} \mid \mathcal{B}_a)\} \ \text{if}\ \tilde{M}_a \ne \emptyset, \\
\hat{B}(aj) &:= \{j = \min\{l \in \tilde{M}_a : E(U_{al} \mid \mathcal{B}_a) = \max_{i \in \tilde{M}_a} E(U_{ai} \mid \mathcal{B}_a)\}\}
\end{aligned}$$

for $\tilde{M}_a \ne \emptyset, j \in \tilde{M}_a$, and

$$\tau_a^*(x) := \begin{cases}
a & \text{if}\ U_a(x) = Z_a(x) \\
b = (a_1, \ldots, a_k, b_1, \ldots, b_j)\ \text{if for}\ b^i \\
\quad U_{b^i}(x) > Z_{b^i}(x),\ x \in \hat{B}(b^{i+1}), k \le i < k+j, \\
\quad U_b(x) = Z_b(x).
\end{cases}$$

Then for all $a \in \tilde{A}$

(i) $\tau_a^\star \in T_a$

(ii) $E(Z_{\tau_a^\star} \mid \mathcal{B}_a) = U_a \quad P - a.s.$

(iii) $U_a \geq E(Z_\tau \mid \mathcal{B}_a) \quad P - a.s. \quad \forall \tau : \tau \text{ sampling plan s.t. } \tau \succeq a.$

(iv) $EZ_{\tau_a^\star} = \sup_{\tau \in T_a} EZ_\tau;$ in particular $EZ_{\tau_{()}^\star} = v.$

Proof: (i) By definition the functions U_a, $a \in \tilde{\mathcal{A}}$, are $(\mathcal{B}_a, \overline{I\!B})$-measurable, and for all $a \in \tilde{\mathcal{A}}$ s.t. $\tilde{M}_a \neq \emptyset$ and all $j \in \tilde{M}_a$ the sets $\hat{B}(aj)$ belong to $\mathcal{B}_a$. Therefore

$$\{\tau_a^\star = b\} \in \mathcal{B}_b \quad \forall b \in \tilde{\mathcal{A}}$$

and

$$\{\tau_a^\star \succeq bj\} \in \mathcal{B}_b \quad \forall b \in \tilde{\mathcal{A}} \text{ s.t. } \tilde{M}_b \neq \emptyset, j \in \tilde{M}_b.$$

According to the definition of Z_a for $a \in \tilde{\mathcal{A}} \backslash \mathcal{A}$ one moreover obtains

$$P(\{\tau_a^\star \in \mathcal{A}\}) = 1 \quad \forall a \in \tilde{\mathcal{A}}$$

and, since $\mathcal{A}$ is finite,

$$EZ_{\tau_a^\star}^- \leq \sum_{b \in \mathcal{A}: b \succeq a} EZ_b^- < \infty,$$

i.e. $\tau_a^\star \in T_a \ \forall a \in \tilde{\mathcal{A}}.$

(ii)/(iii) are shown by backward induction: For each $a \in \tilde{\mathcal{A}}$ s.t. $\tilde{M}_a = \emptyset$

$$E(Z_{\tau_a^\star} \mid \mathcal{B}_a) = Z_a = U_a = E(Z_\tau \mid \mathcal{B}_a) \quad P - a.s.$$

for all sampling plans $\tau \succeq a$, i.e. (ii)/(iii) are fulfilled for these $a \in \tilde{\mathcal{A}}$ – in particular for $a \in \mathcal{A}$ s.t. $g(a) = m := \max_{b \in \mathcal{A}} g(b).$

Consider now $i \in \{1, \ldots, m\}$ and assume that (ii)/(iii) hold for all $a \in \tilde{\mathcal{A}}^j := \{a \in \tilde{\mathcal{A}} : g(a) = j\}$ and all j s.t. $i \leq j \leq m$; let $a \in \tilde{\mathcal{A}}^{i-1}$ and $\tau \succeq a$ be a sampling plan. For each $k \in \tilde{M}_a$

$$\hat{\tau}_k := \begin{cases} \tau & \text{on } \{\tau \succeq ak\} \\ ak & \text{elsewhere} \end{cases}$$

defines a sampling plan $\hat{\tau}_k \succeq ak$. For $C \subset \mathcal{B}_a$ follows

$$\begin{aligned}
\int_C Z_\tau \, dP &= \int_{C \cap \{\tau = a\}} Z_\tau \, dP + \sum_{k \in \tilde{M}_a} \int_{C \cap \{\tau \succeq ak\}} Z_\tau \, dP \\
&= \int_{C \cap \{\tau = a\}} Z_a \, dP + \sum_{k \in \tilde{M}_a} \int_{C \cap \{\tau \succeq ak\}} Z_{\hat{\tau}_k} \, dP \\
&= \int_{C \cap \{\tau = a\}} Z_a \, dP + \sum_{k \in \tilde{M}_a} \int_{C \cap \{\tau \succeq ak\}} E(E(Z_{\hat{\tau}_k} \mid \mathcal{B}_{ak}) \mid \mathcal{B}_a) dP
\end{aligned}$$

$$\text{since } C \cap \{\tau \succeq ak\} \in \mathcal{B}_a \text{ and } \mathcal{B}_a \subset \mathcal{B}_{ak} \forall k \in \tilde{M}_a$$

$$\leq \int_{C \cap \{\tau = a\}} Z_a \, dP + \sum_{k \in \tilde{M}_a} \int_{C \cap \{\tau \succeq ak\}} E(U_{ak} \mid \mathcal{B}_a) \, dP$$

by assumption (part (iii)), where by assumption (part (ii)) equality holds for $\tau = \tau_a^\star$

$$\leq \int_C \max(Z_a, \max_{k \in \tilde{M}_a} E(U_{ak} \mid \mathcal{B}_a))dP$$

where, according to the construction, equality holds for $\tau = \tau_a^\star$

$$= \int_C U_a \, dP$$

by the definition of U_a.

Hence it follows

$$E(Z_\tau \mid \mathcal{B}_a) \leq U_a \quad P - a.s.$$

for each sampling plan $\tau \succeq a$ and

$$E(Z_{\tau_a^\star} \mid \mathcal{B}_a) = U_a \quad P - a.s.;$$

this yields assertions (ii) and (iii).

(iv) From (ii)/(iii) one obtains by integration

$$EZ_{\tau_a^\star} \overset{(ii)}{=} EU_a \overset{(iii)}{\geq} EZ_\tau \quad \forall \tau \in T_a$$

and therefore the desired result. $\qquad\qquad\square$

(2.6) Remark

The basic idea for the construction of $\tau_a^\star$ is to start with a fixed sample sequence $a \in \tilde{\mathcal{A}}$ and to continue, for $b \in \tilde{\mathcal{A}}$, in case of $U_b(x) > Z_b(x)$ the course of observation in an optimal way by choosing the size of the next sub-sample according to $x \in \hat{B}(bk)$ and to stop sampling in case of $U_b(x) = Z_b(x)$ (where the pay-off to be expected under the "information" $\mathcal{B}_a$ is already maximal). The definition

$$Z_b(x) := -\infty \quad \forall b \in \tilde{\mathcal{A}} \backslash \mathcal{A}$$

has the consequence that the observations are stopped with admissible sample sequences $a \in \mathcal{A}$ only. Assertion (2.5) (iv) gives – at least in principle – a complete solution for problems of optimal sampling with finite horizon: $\tau_{()}^\star \in T_{()} = T$ is an *optimal* sequential sampling plan. $\qquad\qquad\square$

The third variant of example (2.3) leads to such a problem of optimal sampling with finite $\mathcal{A}$. For an illustration we solve a simple special case:

(2.7) Example

Consider the modified S_n/n-problem for $m = 6$. By a direct application of the construction method of (2.5) one can compute, for all $\mid \mathcal{A} \mid = 64$ admissible sample sequences $a \in \mathcal{A}$, the functions U_a and the corresponding sampling plans $\tau_a^\star$.

$$\tau_{()}^\star(x_1, \ldots, x_6) := \begin{cases} (3) & \text{if } \sum_{i=1}^{3} x_i > 0 \\ (3,1,1) & \text{if } \sum_{i=1}^{3} x_i = -1, \sum_{i=1}^{5} x_i = 1 \\ (3,1,1,1) & \text{elsewhere} \end{cases}$$

turns out to be an optimal sequential sampling plan with expected pay-off $v = EZ_{\tau_{()}^*} = 73/128$. Using this plan one starts with an initial sample of size 3. If $\sum_{i=1}^{3} x_i = 1$ or 3 sampling is stopped; else sampling is continued by single observations. In case of $\sum_{i=1}^{5} x_i = 1$ (which is possible for $\sum_{i=1}^{3} x_i = -1$ only) no further observation is made; else one goes on up to the maximal total sample size 6. The value $v = 0,5703125$ is considerably higher than the value of the corresponding finite S_n/n-problem (which is $0,4395833$).

§ 3 Existence of optimal sampling plans for general $\mathcal{A}$

Our next goal is to prove (under mild additional assumptions) assertions on the existence and structure of optimal sampling plans for arbitrary sets of admissible sample sequences. As in the theory of optimal stopping we always assume that the stochastic process $(Z_a)_{a\in\mathcal{A}}$ fulfills[23]

$$E(\sup_{a\in\mathcal{A}} Z_a^+) < \infty.$$

In § 2 obviously

$$U_a = \operatorname*{ess\,sup}_{\tau\in T_a} E(Z_\tau \mid \mathcal{B}_a) \quad P-a.s.\ \forall a \in \tilde{A}.$$

Therefore the question arises whether corresponding terms lead to optimal sampling plans also in the general case. Let therefore

$$\tilde{A} := \{\tilde{a} \in A : \exists a \in \mathcal{A}\ \text{s.t.}\ \tilde{a} \preceq a\}$$

again be the minimal sub-tree of A whose smallest element is () and which contains $\mathcal{A}$, and

$$\begin{aligned}
\tilde{M}_a &:= \{j \in M : aj \in \tilde{A}\},\ a \in \tilde{A}, \\
U_a &:= \operatorname*{ess\,sup}_{\tau\in T_a} E(Z_\tau \mid \mathcal{B}_a),\ a \in \tilde{A}.
\end{aligned}$$

It should be mentioned that each $U_a, a \in \tilde{A}$ is a $(\mathcal{B}_a, \overline{I\!\!B})$-measurable random variable which is P-a.s. unique and because of

$$E\,U_a \leq E(E(\sup_{b\in\mathcal{A}} Z_b^+ \mid \mathcal{B}_a)) = E(\sup_{b\in\mathcal{A}} Z_b^+) < \infty$$

quasiintegrable. Moreover, for U_a one obtains a representation similar to that of (2.5):

(2.8) Lemma *(Bellman-Equation)*
  For each $a \in \tilde{A}$ holds[24]

$$U_a = \max\{Z_a, \sup_{k\in\tilde{M}_a} E(U_{ak} \mid \mathcal{B}_a)\} \quad P-a.s.$$

(Without loss of generality we assume this relation to hold for all $x \in \mathcal{X}$.)
Proof: We show that for each $a \in \tilde{A}$

---

[23]This assumption is e.g. fulfilled if $Z_a \leq 0$ (i.e. allow an interpretation as losses); comp. Chapter 4.
[24]We always define $\sup_{\emptyset} \ldots = \max_{\emptyset} \ldots = -\infty$.

36

(i) $U_a = \max\{Z_a, \operatorname*{ess\,sup}_{\tau \in T_{a+}} E(Z_\tau \mid \mathcal{B}_a)\}\ P - a.s.,$

(ii) $\operatorname*{ess\,sup}_{\tau \in T_{ak}} E(Z_\tau \mid \mathcal{B}_a) = E(U_{ak} \mid \mathcal{B}_a)\ P - a.s.\ \forall k \in \tilde{M}_a$ (if $\tilde{M}_a \neq \emptyset$),

(iii) $\operatorname*{ess\,sup}_{\tau \in T_{a+}} E(Z_\tau \mid \mathcal{B}_a) = \sup_{k \in \tilde{M}_a} E(U_{ak} \mid \mathcal{B}_a)\ P - a.s..$

Then the assertion of (2.8) follows from (i) and (iii).

<u>(i)</u>: For $a \in \tilde{\mathcal{A}}$ s.t. $\tilde{M}_a = \emptyset$ part (i) is obviously fulfilled. Therefore, only $a \in \tilde{\mathcal{A}}$ s.t. $\tilde{M}_a \neq \emptyset$ are considered; for fixed $a \in \tilde{\mathcal{A}}$ we use the abbreviation

$$V := \operatorname*{ess\,sup}_{\tau \in T_{a+}} E(Z_\tau \mid \mathcal{B}_a).$$

For the constant sampling plan $\tau \equiv a$

$$U_a \geq E(Z_\tau \mid \mathcal{B}_a) = Z_a\ \ P - a.s. \text{ if } a \in \mathcal{A}$$

and

$$U_a \geq -\infty = Z_a\ \ \ P - a.s. \text{ if } a \notin \mathcal{A}.$$

Moreover, from $T_{a+} \subset T_a$ follows

$$U_a = \operatorname*{ess\,sup}_{\tau \in T_a} E(Z_\tau \mid \mathcal{B}_a) \geq \operatorname*{ess\,sup}_{\tau \in T_{a+}} E(Z_\tau \mid \mathcal{B}_a) = V\ \ P - a.s.,$$

i.e. $U_a \geq \max\{Z_a, V\}\ P$-a.s.. Defining on the other hand for an arbitrary $\tau \in T_a$ a further sampling plan $\tilde{\tau} \in T_{a+}$ by

$$\tilde{\tau} := \begin{cases} \tau & \text{on} \quad \{\tau \succ a\} \\ b & \text{on} \quad \{\tau = a\} \end{cases}$$

where b is a fixed admissible sample sequence $\succeq a$ one obtains

$$\begin{aligned}
E(Z_\tau \mid \mathcal{B}_a) &= E(1_{\{\tau=a\}} \cdot Z_a \mid \mathcal{B}_a) + E(1_{\{\tau \succ a\}} \cdot Z_{\tilde{\tau}} \mid \mathcal{B}_a) \\
&\leq 1_{\{\tau=a\}} \cdot Z_a + 1_{\{\tau \succ a\}} \operatorname*{ess\,sup}_{\hat{\tau} \in T_a^+} E(Z_{\hat{\tau}} \mid \mathcal{B}_a) \\
&= 1_{\{\tau=a\}} \cdot Z_a + 1_{\{\tau \succ a\}} \cdot V \\
&\leq \max\{Z_a, V\}\ \ P - a.s..
\end{aligned}$$

Hence it follows

$$U_a = \operatorname*{ess\,sup}_{\tau \in T_a} E(Z_\tau \mid \mathcal{B}_a) \leq \max\{Z_a, V\}\ \ P - a.s.,$$

i.e. assertion (i).

<u>(ii)</u>: Consider $a \in \tilde{\mathcal{A}}$ s.t. $\tilde{M}_a \neq \emptyset$ and $k \in \tilde{M}_a$. Then

$$E(U_{ak} \mid \mathcal{B}_a) \geq E(E(Z_\tau \mid \mathcal{B}_{ak}) \mid \mathcal{B}_a) = E(Z_\tau \mid \mathcal{B}_a)\ P - a.s.$$

for all $\tau \in T_{ak}$ and therefore

$$\operatorname*{ess\,sup}_{\tau \in T_{ak}} E(Z_\tau \mid \mathcal{B}_a) \leq E(U_{ak} \mid \mathcal{B}_a) \quad P - a.s..$$

On the other hand one obtains from the properties of the essential supremum (see e.g. [Ir], Appendix A.2) that there exists a sequence $(\tau_n)_{n \in \mathbb{N}}$ of sampling plans $\tau_n \in T_{ak}$ such that

$$U_{ak} = \sup_{n \in \mathbb{N}} E(Z_{\tau_n} \mid \mathcal{B}_{ak}) \quad P - a.s..$$

Without loss of generality we assume $\tau_1 \equiv b$ where $b \in \mathcal{A}, b \succeq ak$ is fixed. Defining

$$\tilde{\tau}_1 := \tau_1$$
$$\tilde{\tau}_n := \begin{cases} \tilde{\tau}_{n-1} & \text{on} \quad B_n \\ \\ \tau_n & \text{on} \quad B_n^c \end{cases} , n \geq 2,$$

where

$$B_n := \{ E(Z_{\tilde{\tau}_{n-1}} \mid \mathcal{B}_{ak}) \geq E(Z_{\tau_n} \mid \mathcal{B}_{ak}) \}$$

yields a further sequence $(\tilde{\tau}_n)_{n \in \mathbb{N}}$ of sampling plans $\tilde{\tau}_n \in T_{ak}$. According to the definition of B_n it follows for these plans

$$\begin{aligned} E(Z_{\tilde{\tau}_n} \mid \mathcal{B}_{ak}) &= E(1_{B_n} \cdot Z_{\tilde{\tau}_{n-1}} \mid \mathcal{B}_{ak}) + E(1_{B_n^c} \cdot Z_{\tau_n} \mid \mathcal{B}_{ak}) \\ &= 1_{B_n} \cdot E(Z_{\tilde{\tau}_{n-1}} \mid \mathcal{B}_{ak}) + 1_{B_n^c} \cdot E(Z_{\tau_n} \mid \mathcal{B}_{ak}) \\ &= \max\{ E(Z_{\tilde{\tau}_{n-1}} \mid \mathcal{B}_{ak}), \ E(Z_{\tau_n} \mid \mathcal{B}_{ak}) \} \ P - a.s. \end{aligned}$$

for all $n > 1$. Therefore the sequence

$$(E(Z_{\tilde{\tau}_n} \mid \mathcal{B}_{ak}))_{n \in \mathbb{N}}$$

is P-a.s. non-decreasing, and for all $n \in \mathbb{N}$

$$E(Z_{\tilde{\tau}_n} \mid \mathcal{B}_{ak}) \geq E(Z_{\tau_n} \mid \mathcal{B}_{ak}) \ P - a.s..$$

Hence

$$\begin{aligned} U_{ak} &= \sup_{n \in \mathbb{N}} E(Z_{\tau_n} \mid \mathcal{B}_{ak}) \leq \sup_{n \in \mathbb{N}} E(Z_{\tilde{\tau}_n} \mid \mathcal{B}_a) \\ &= \lim_n E(Z_{\tilde{\tau}_n} \mid \mathcal{B}_{ak}) \quad P - a.s. \end{aligned}$$

and, furthermore, according to the monotone convergence theorem for conditional expectations

$$\begin{aligned} E(U_{ak} \mid \mathcal{B}_a) &\leq E(\lim_n E(Z_{\tilde{\tau}_n} \mid \mathcal{B}_{ak}) \mid \mathcal{B}_a) = \\ &= \lim_n E(Z_{\tilde{\tau}_n} \mid \mathcal{B}_a) \leq \operatorname*{ess\,sup}_{\tau \in T_{ak}} E(Z_\tau \mid \mathcal{B}_a) \ P - a.s. \end{aligned}$$

since $\tilde{\tau}_n \in T_{ak} \ \forall n \in \mathbb{N}$; i.e. assertion (ii).

(iii) For $a \in \tilde{A}$ s.t. $\tilde{M}_a \neq \emptyset$ part (iii) is obviously fulfilled; consider therefore $a \in \tilde{A}$ s.t. $\tilde{M}_a \neq \emptyset$. Then it follows for all $k \in \tilde{M}_a$

$$\operatorname*{ess\,sup}_{\tau \in T_{a+}} E(Z_\tau \mid \mathcal{B}_a) \geq \operatorname*{ess\,sup}_{\tau \in T_{ak}} E(Z_\tau \mid \mathcal{B}_a), \text{ since } T_{ak} \subset T_{a+}$$
$$= E(U_{ak} \mid \mathcal{B}_a) \; P - a.s. \text{ according to (ii)}$$

and therefore

$$\operatorname*{ess\,sup}_{\tau \in T_{a+}} E(Z_\tau \mid \mathcal{B}_a) \geq \sup_{k \in \tilde{M}_a} E(U_{ak} \mid \mathcal{B}_a) \; P - a.s..$$

Defining on the other hand, for a given sampling plan $\tau \in T_{a+}$, for each $k \in \tilde{M}_a$ a sampling plan $\tau_k \in T_{ak}$ by

$$\tau_k := \begin{cases} \tau & \text{on } \{\tau \succeq ak\} \\ ak & \text{elsewhere} \end{cases}$$

one obtains

$$
\begin{aligned}
E(Z_\tau \mid \mathcal{B}_a) &= E\Big(\sum_{k \in \tilde{M}_a} 1_{\{\tau \succeq ak\}} \cdot Z_{\tau_k} \mid \mathcal{B}_a \Big) \\
&= E\Big(\sum_{k \in \tilde{M}_a} 1_{\{\tau \succeq ak\}} \cdot (Z_{\tau_k})^+ \mid \mathcal{B}_a \Big) - E\Big(\sum_{k \in \tilde{M}_a} 1_{\{\tau \succeq ak\}} \cdot (Z_{\tau_k})^- \mid \mathcal{B}_a \Big) \\
&= \sum_{k \in \tilde{M}_a} 1_{\{\tau \succeq ak\}} \cdot E(Z_{\tau_k} \mid \mathcal{B}_a) \\
&\quad \text{(monotone convergence theorem for conditional expectations} \\
&\quad \text{applied to both terms)} \\
&= \sum_{k \in \tilde{M}_a} 1_{\{\tau \succeq ak\}} \cdot E\big(E(Z_{\tau_k} \mid \mathcal{B}_{ak}) \mid \mathcal{B}_a\big) \\
&\leq \sum_{k \in \tilde{M}_a} 1_{\{\tau \succeq ak\}} \cdot E(U_{ak} \mid \mathcal{B}_a) \text{ since } \tau_k \in T_{ak} \; \forall k \in \tilde{M}_a \\
&\leq \sup_{k \in \tilde{M}_a} E(U_{ak} \mid \mathcal{B}_a) \; P - a.s..
\end{aligned}
$$

This yields

$$\operatorname*{ess\,sup}_{\tau \in T_{a+}} E(Z_\tau \mid \mathcal{B}_a) \leq \sup_{k \in \tilde{M}_a} E(U_{ak} \mid \mathcal{B}_a) \; P - a.s.$$

and altogether assertion (iii). $\qquad\square$

(2.9) Remark

Trying to define a sampling plan analogously to $\tau_0^\star$ in (2.6) one is faced, for $|M| = \infty$, with the problem that there might exist $a \in \tilde{A}$ and $x \in \mathcal{X}$ for which the *supremum* in Bellman's equation is *not realized*, i.e. that

$$B(a\,\infty) := \Big\{ E(U_{ak} \mid \mathcal{B}_a) < \sup_{j \in \tilde{M}_a} E(U_{aj} \mid \mathcal{B}_a) \; \forall k \in \tilde{M}_a \Big\} \cap \{U_a > Z_a\}$$

is non-empty. This is taken into account at the following construction: Define for $a \in \tilde{A}$ s.t. $\tilde{M}_a \neq \emptyset$ and $j \in \tilde{M}_a$ in analogy to § 2

$$B(aj) := \Big\{ j = \min\{k \in \tilde{M}_a : E(U_{ak} \mid \mathcal{B}_a) = \sup_{i \in \tilde{M}_a} E(U_{ai} \mid \mathcal{B}_a)\} \Big\} \cap \{U_a > Z_a\}$$

and

$$\tau^\star(x) := \begin{cases} () & \text{if } U_{()}(x) = Z_{()}(x) \\ a = (a_1, \ldots, a_j) & \text{if } x \in B(a^{i+1}), 0 \le i < j, U_a(x) = Z_a(x) \\ a = (a_1, a_2, \ldots) & \text{if } x \in B(a^{i+1}) \; \forall i \in \mathbb{N}_0 \\ a = (a_1, \ldots, a_j, m, m, \ldots) & \text{where } m := \min M, \text{if } x \in B(a^{i+1}), \\ & \quad 0 \le i < j, x \in B(a^j \infty). \end{cases}$$

Then

$$\{\tau^\star = a\} \in \mathcal{B}_a$$

for all $a = (a_1, \ldots, a_j) \in \tilde{A}$ since $B(a^{i+1}) \in \mathcal{B}_{a^i} \subset \mathcal{B}_a \; \forall i : 0 \le i < j$ and since U_a and Z_a are $(\mathcal{B}_a, \mathbb{B})$-measurable functions; for $a \in A \backslash \tilde{A}$ follows

$$\{\tau^\star = a\} = \emptyset \quad (\in \mathcal{B}_a).$$

In a similar way one obtains for all $a = (a_1, \ldots, a_j) \in A$ and $k \in M$

$$\{\tau^\star \succeq ak\} = \begin{cases} \cap_{i=0}^{j-1} B(a^{i+1}) \cap B(ak) + \cup_{i=r}^{j} \cap_{l=0}^{i-1} B(a^{l+1}) \cap B(a^i \infty) \\ \quad \in \mathcal{B}_a \text{ if } k = m, \; r := \max\{s : a_s \ne m\} \\ \cap_{i=0}^{j-1} B(a^{i+1}) \cap B(ak) \in \mathcal{B}_a \text{ if } k \ne m; \end{cases}$$

i.e. the measurability conditions (of (1.22)) on a sequential sampling plan are fulfilled.

Similarly as $\tau_{()}^\star$ in (2.6), $\tau^\star$ starts sampling at () and stops at $a \in \tilde{A}$ if $U_a(x) = Z_a(x)$, i.e. if the maximal pay-off $U_a(x)$ to be expected under the "information" $\mathcal{B}_a$ is already achieved. Due to $Z_a \equiv -\infty \; \forall a \in A \backslash \mathcal{A}$ this condition to stop will P-a.s. be fulfilled only for sample sequences $a \in \mathcal{A}$, i.e.

$$P(\{\tau^\star \in A \backslash \mathcal{A}\}) = 0.$$

If $U_a(x) > Z_a(x)$ the size of the next sub-sample is *minimal* among the optimal continuations – provided that the supremum

$$\sup_{i \in \tilde{M}_a} E(U_{ai} \mid \mathcal{B}_a)$$

can be realized. If this supremum is not achieved the further course of sampling is (arbitrarily) defined by always making m additional observations – sampling will *never stop* in this case.

The following optimality result shows that considering $\tau^\star$ may be useful:

(2.10) Theorem *(see also Haggstrom [Ha]; Theorem 4.1)*

If $P(\{\tau^\star \in A\}) = 1$ then $\tau^\star$ is an optimal sampling plan.

Proof: Because of the measurability properties of $\tau^\star$ and the assumption $P(\{\tau^\star \in A\}) = 1$ it suffices to show that

$$(\star) \qquad U_{()} = E(Z_{\tau^\star} \mid \mathcal{B}_{()}) \quad P - a.s.;$$

40

then by the definition of $U_{()}$ follows

$$E\,Z_{\tau^*} = EU_{()} \geq E(E(Z_\tau \mid \mathcal{B}_{()})) = E\,Z_\tau \quad \forall \tau \in T,$$

i.e. $\tau^* \in T$ as well as $E\,Z_{\tau^*} = v$.

For that purpose we prove that for all $n \in \mathbb{N}_0$

$(\star\star) \quad U_{()} = E(1_{\{\tau^* \prec \tilde{\mathcal{A}}_n\}} \cdot Z_{\tau^*} + \sum_{a \in \tilde{\mathcal{A}}_n} 1_{\{\tau^* \succeq a\}} \cdot U_a \mid \mathcal{B}_a)\ P - a.s.$

where $\tilde{\mathcal{A}}_n := \{a \in \tilde{\mathcal{A}} : h(a) = n\}$ and $\{\tau^* \prec \tilde{\mathcal{A}}_n\} = \cup_{j=0}^{n-1} \cup_{a \in \tilde{\mathcal{A}}_j} \{\tau^* = a\}$, in particular $\{\tau^* \prec \tilde{\mathcal{A}}_0\} = \emptyset$.

$(\star\star)$ is shown by induction (on n): For $n = 0$ we obtain

$$E(1_{\{\tau^* \prec \tilde{\mathcal{A}}_0\}} \cdot Z_{\tau^*} \;+\; \sum_{a \in \tilde{\mathcal{A}}_0} 1_{\{\tau^* \succeq a\}} U_a \mid \mathcal{B}_{()}) =$$
$$= E(1_{\{\tau^* \succeq ()\}} \cdot U_{()} \mid \mathcal{B}_{()}) = U_{()} \quad P - a.s..$$

Assume that $(\star\star)$ holds for $n \in \mathbb{N}_0$; then one obtains, due to the properties of U_a and the construction of τ^*, for the second term

$$E(\sum_{a \in \tilde{\mathcal{A}}_n} 1_{\{\tau^* \succeq a\}} \cdot U_a \mid \mathcal{B}_{()}) =$$

$$= E(\sum_{a \in \tilde{\mathcal{A}}_n} 1_{\{\tau^* = a\}} \cdot Z_a + \sum_{a \in \tilde{\mathcal{A}}_n} \sum_{j \in \tilde{M}_a} 1_{\{\tau^* \succeq aj\} \cap \bigcap_{b \prec a} B(b\infty)^c} \cdot E(U_{aj} \mid \mathcal{B}_a)$$
$$+ \sum_{a \in \tilde{\mathcal{A}}_n} U_a \cdot 1_{\{\tau^* \succeq am\} \cap \bigcup_{b \prec a} B(b\infty)} \mid \mathcal{B}_{()})$$

$$= E(1_{\{\tau^* \in \tilde{\mathcal{A}}_n\}} \cdot Z_{\tau^*} + \sum_{a \in \tilde{\mathcal{A}}_n} \sum_{j \in \tilde{M}_a} 1_{\{\tau^* \succeq aj\}} \cdot E(U_{aj} \mid \mathcal{B}_a) \mid \mathcal{B}_{()})$$
$$(\text{since } P(\{\tau^* \in \mathcal{A}\}) = 1)$$

$$= E(1_{\{\tau^* \in \tilde{\mathcal{A}}_n\}} \cdot Z_{\tau^*} \mid \mathcal{B}_{()}) + E(\sum_{a \in \tilde{\mathcal{A}}_n} \sum_{j \in \tilde{M}_a} 1_{\{\tau^* \succeq aj\}} \cdot E(U_{aj}^+ \mid \mathcal{B}_a) \mid \mathcal{B}_{()})$$
$$- E(\sum_{a \in \tilde{\mathcal{A}}_n} \sum_{j \in \tilde{M}_a} 1_{\{\tau^* \succeq aj\}} \cdot E(U_{aj}^- \mid \mathcal{B}_a) \mid \mathcal{B}_{()})$$

$$= E(1_{\{\tau^* \in \tilde{\mathcal{A}}_n\}} \cdot Z_{\tau^*} \mid \mathcal{B}_{()}) + \sum_{a \in \tilde{\mathcal{A}}_n} \sum_{j \in \tilde{M}_a} E(1_{\{\tau^* \succeq aj\}} \cdot U_{aj} \mid \mathcal{B}_{()})$$
$$(\text{monotone convergence theorem for conditional expectations})$$

$$= E(1_{\{\tau^* \in \tilde{\mathcal{A}}_n\}} \cdot Z_{\tau^*} + \sum_{b \in \tilde{\mathcal{A}}_{n+1}} 1_{\{\tau^* \succeq b\}} \cdot U_b \mid \mathcal{B}_{()})\ P - a.s.$$
$$(\text{again due to monotone convergence}).$$

Hence it follows

$$U_{()} = E(1_{\{\tau^* \prec \tilde{\mathcal{A}}_n\}} \cdot Z_{\tau^*} \mid \mathcal{B}_{()}) + E(1_{\{\tau^* \in \tilde{\mathcal{A}}_n\}} \cdot Z_{\tau^*} + \sum_{b \in \tilde{\mathcal{A}}_{n+1}} 1_{\{\tau^* \succeq b\}} \cdot U_b \mid \mathcal{B}_{()})$$

$$= E(1_{\{\tau^* \prec \tilde{\mathcal{A}}_{n+1}\}} \cdot Z_{\tau^*} + \sum_{b \in \tilde{\mathcal{A}}_{n+1}} 1_{\{\tau^* \succeq b\}} \cdot U_b \mid \mathcal{B}_{()})\ P - a.s.,$$

i.e. $(\star\star)$ holds for $n + 1$, too.

From $\tau^* \in \mathcal{A}$ P-a.s. and $E(\sup_{b \in \mathcal{A}} Z_b^+) < \infty$ follows

$$\limsup_{n \to \infty} E(\textstyle\sum_{a \in \tilde{A}_n} 1_{\{\tau^* \succeq a\}} \cdot U_a \mid \mathcal{B}_{()}) \leq$$
$$\limsup_{n \to \infty} E(1_{\{\tau^* \succeq \tilde{A}_n\}} \cdot \sup_{b \in \mathcal{A}} Z_b^+ \mid \mathcal{B}_{()}) = 0 \; P - a.s..$$

$(\star\star)$ therefore yields

$$
\begin{aligned}
U_{()} \;&=\; \limsup_{n \to \infty} E(1_{\{\tau^* \prec \tilde{A}_n\}} \cdot Z_{\tau^*} + \sum_{a \in \tilde{A}_n} 1_{\{\tau^* \succeq a\}} \cdot U_a \mid \mathcal{B}_{()}) \\
&\leq\; \limsup_{n \to \infty} E(1_{\{\tau^* \prec \tilde{A}_n\}} \cdot Z_{\tau^*} \mid \mathcal{B}_{()}) \; P - a.s..
\end{aligned}
$$

Because of $1_{\{\tau^* \prec \tilde{A}_n\}} \uparrow 1$ P-a.s. one therefore obtains (again due to monotone convergence)

$$
\begin{aligned}
U_{()} \;&\leq\; \limsup_{n \to \infty} E(1_{\{\tau^* \prec \tilde{A}_n\}} \cdot Z_{\tau^*}^+ \mid \mathcal{B}_{()}) - \liminf_{n \to \infty} E(1_{\{\tau^* \prec \tilde{A}_n\}} \cdot Z_{\tau^*}^- \mid \mathcal{B}_{()}) \\
&=\; E(Z_{\tau^*} \mid \mathcal{B}_{()}) \; P - a.s.,
\end{aligned}
$$

i.e. a first part of $(\star)$. In particular it follows

$$-\infty < E\, U_{()} \leq E\, Z_{\tau^*}$$

and therefore $E(Z_{\tau^*}^-) < \infty$, i.e. $\tau^* \in T$. From

$$U_{()} = \operatorname*{ess\,sup}_{\tau \in T} E(Z_\tau \mid \mathcal{B}_{()}) \geq E(Z_{\tau^*} \mid \mathcal{B}_{()}) \; P - a.s.$$

then follows the remaining part of $(\star)$. $\qquad\square$

(2.11) Remark

According to (2.10) the condition $P(\{\tau^* \in \mathcal{A}\}) = 1$ is *sufficient* for the optimality of τ^*. It turns out that the property

$$(+) \qquad P(\{\tau^* \succ a\} \cap B(a\infty)) = 0 \quad \forall a \in \tilde{\mathcal{A}},$$

resulting from that condition, is *necessary* for the optimality of τ^*: If condition $(+)$ is violated then the $\mathcal{B}$-measurable set

$$D := \bigcup_{a \in \tilde{\mathcal{A}}} \{\tau^* \succ a\} \cap B(a\infty)$$

has positive probability. But $\tau^* \in A^*$ on D and therefore $Z_{\tau^*} = -\infty$ on D. Since in

$$EZ_{\tau^*} = \int_D Z_{\tau^*} \, dP + \int_{D^c} Z_{\tau^*} dP \leq \int_D Z_{\tau^*} dP + E(\sup_{a \in \mathcal{A}} Z_a^+)$$

the last term is finite by assumption it follows

$$EZ_{\tau^*} = -\infty < EZ_\tau \quad \forall \tau \in T,$$

i.e. τ^* yields the least favourable expected pay-off. Condition $(+)$ is, in particular, always fulfilled if M is *finite*, i.e. if an upper bound for the size of the sub-samples exists – then also all sets $\tilde{M}_a$ are finite and the $\sup_{j \in \tilde{M}_a} E(U_{aj} \mid \mathcal{B}_a)$ are realized. $\qquad\square$

42

Up to now our assertions were suggested by corresponding results of the theory of optimal stopping; the next (deterministic) example shows limitations of these analogies:

(2.12) Example
Consider

$$M = \{1,2\}, \ \mathcal{A} = A = \cup_{j \in \mathbb{N}} M^j \cup \{()\}, \ Z_{()} = 0$$

and

$$Z_{(a_1,\dots,a_j)} := \begin{cases} 0 & \text{if } a_j = 1 \\ 1 & \text{elsewhere.} \end{cases}$$

Then obviously

$$v = \sup_{\tau \in T} E Z_\tau = 1$$

and v is (e.g.) realized by the simple sampling plan $\hat{\tau} = (2) \in T$. On the other hand one obtains for each $a \in \mathcal{A}$

$$B(a1) = \begin{cases} \mathcal{X} & \text{if } a = () \text{ or if } a_{h(a)} = 1 \\ \emptyset & \text{elsewhere} \end{cases}$$

and

$$B(a2) = \emptyset = B(a\infty);$$

hence

$$\tau^* = (1, 1, \dots),$$

i.e. τ^* is not even a sampling plan. Contrary to the theory of optimal stopping (see [C/R/S], Theorem 4.2) the existence of an optimal sampling plan does i.g. *not yield* the optimality of τ^*. $\qquad\square$

For an illustration of our previous results we again come back to example (2.3):

(2.13) Examples
a) For the modified S_n/n-problem (2.3) (ii) the crude estimation

$$Z_{(a_1,\dots,a_j)} = \frac{1}{j} \sum_{i=1}^{a_1+\dots+a_j} X_j \leq m$$

shows that the general condition $E(\sup_{a \in \mathcal{A}} Z_a^+) < \infty$ is fulfilled. In fact, one can show that there exists an optimal sampling plan for this problem i.e. a $\tilde{\tau} \in T$ such that

$$E Z_{\tilde{\tau}} = \sup_{\tau \in T} E Z_\tau :$$

The basic idea is that one may restrict attention on sampling plans where each sub-sample has maximal size m and that therefore results of the "classical" $S_n/n-$problem may be used. For $a = (a_1, \dots, a_j) \in A$ define

$$\mathcal{A}_a := \{a \times (m)_{1 \leq i \leq k} := (a_1, \dots, a_j, \underbrace{m, \dots, m}_{k\text{-times}}) : k \in \mathbb{N}_0\}$$

and

$$T_a^{(m)} := \{\tau \in T : P(\tau \in \mathcal{A}_a) = 1\}.$$

Assertion: For each $a \in A$

$$\operatorname*{ess\,sup}_{\tau \in T_a} E(Z_\tau \mid \mathcal{B}_a) = \operatorname*{ess\,sup}_{\tau \in T_a^{(m)}} E(Z_\tau \mid \mathcal{B}_a).$$

Proof: Let $a \in A$ be arbitrary. From $T_a^{(m)} \subset T_a$ the " $\geq$ "-relation follows. To show the remaining inequality define for $\tau \in T_a$

$$\hat{\tau} := \begin{cases} a & \text{if } \tau = a \\ a \times (m)_{1 \leq i \leq k} & \text{if } \tau \in A_{h(a)+k}, k \in \mathbb{N} \\ a \times (m)_{i \in \mathbb{N}} & \text{if } \tau \in A^\star. \end{cases}$$

Then $\hat{\tau} \in T_a^{(m)}$ and

$$
\begin{aligned}
E(Z_{\hat{\tau}} \mid \mathcal{B}_a) &= E\Big(\sum_{n=h(a)}^{\infty} \sum_{\substack{j=n \\ j \geq g(a)}}^{n \cdot m} 1_{\{h \circ \tau = n, g \circ \tau = j\}} \frac{S_{g \circ \hat{\tau}}}{h \circ \hat{\tau}} \mid \mathcal{B}_a \Big) \\
&= E\Big(\sum_{n=h(a)}^{\infty} \sum_{\substack{j=n \\ j \geq g(a)}}^{n \cdot m} 1_{\{h \circ \tau = n, g \circ \tau = j\}} \frac{S_{g \circ \tau} + \sum_{i=g \circ \tau + 1}^{g \circ \hat{\tau}} X_i}{h \circ \tau} \mid \mathcal{B}_a \Big) \\
&\qquad \text{since } h \circ \tau = h \circ \hat{\tau} \\
&= E(Z_\tau \mid \mathcal{B}_a) \\
&\quad + \sum_{n=h(a)}^{\infty} \sum_{\substack{j=n \\ j \geq g(a)}}^{n \cdot m} E\big(1_{\{\tau \in A_{(n)} \cap A^{(j)}\}} E\big(\sum_{i=j+1}^{q} \frac{X_i}{n} \mid \sigma(X_1, \ldots, X_j)\big) \mid \mathcal{B}_a\big), \\
&\qquad q := g(a) + (n - h(a)) \cdot m, \\
&= E(Z_\tau \mid \mathcal{B}_a) + \sum_{n=h(a)}^{\infty} \sum_{\substack{j=n \\ j \geq g(a)}}^{n \cdot m} E\big(1_{\{\tau \in A_{(n)} \cap A^{(j)}\}} \cdot E\big(\sum_{i=j+1}^{q} \frac{X_i}{n}\big) \mid \mathcal{B}_a\big) \\
&\qquad \text{since } \sum_{i=j+1}^{q} X_i/n \text{ and } X_1, \ldots, X_j \text{ are independent} \\
&= E(Z_\tau \mid \mathcal{B}_a) \quad P\text{-a.s., since the } X_i \text{ are contored;}
\end{aligned}
$$

i.e. for each $\tau \in T_a$ there exists a $\hat{\tau} \in T_a^{(m)}$ s.t.

$$E(Z_{\hat{\tau}} \mid \mathcal{B}_a) \geq E(Z_\tau \mid \mathcal{B}_a) \; P\text{-a.s.}.$$

Therefore attention may be restricted on $\tau \in T_{()}^{(m)}$ i.e. sampling plans with fixed sample sizes m. But now theorem (4.11) of [C/R/S] yields the existence of an optimal "sampling plan" $\tilde{\tau} \in T_{()}^{(m)} \subset T$.

b) From the special case of optimal stopping it is well-known that without the usual assumption

$$E(\sup_{a \in A} Z_a^+) < \infty$$

no general proposition on the existence of optimal sampling plans and in particular on the optimality of $\tau^\star$ can be made. Example (2.3) (i) emphasizes this phenomenon once

44

more: It can be shown ([Sch], p. 66-68) that the value of the problem (2.3) (i) is $+\infty$ and that there exists an optimal sampling plan $\hat{\tau}$ with $Z_{\hat{\tau}} \neq U_\tau = \infty$ P-a.s. $\square$

The next proposition gives a *sufficient* condition for the optimality of τ^* which is easy to be verified in particular in statistical applications where costs of sampling and losses due to erroneous decisions occur:

(2.14) Theorem *(comp. [Ha], Theorem 4.3)*

> *Let $P(B(a\infty)) = 0$ for all $a \in \tilde{A}$ and let X be an integrable random variable and $(r_n)_{n \in \mathbb{N}_0}$ a sequence of real numbers s.t. $\lim_{n \to \infty} r_n = -\infty$ and*
>
> $$Z_a \leq X + r_n \quad P\text{-}a.s.$$
>
> *for all $n \in \mathbb{N}_0$ and $a \in \mathcal{A}_n$. Then τ^* is an optimal sampling plan.*

Proof: We show that $\tau^* \in \mathcal{A}$ P-a.s.; then theorem (2.10) yields the desired results. Since $P(B(a\infty)) = 0 \; \forall a \in \tilde{A}$, the functions U_a and $E(U_{ak} \mid \mathcal{B}_a)$ may without any restriction be chosen such that $B(a\infty) = \emptyset$ and therefore

$$\{\tau^* \succeq a\} \cap \bigcup_{b \prec a} B(b\infty) = \emptyset, \quad \{\tau^* \succeq a\} \cap \bigcap_{b \prec a} B(b\infty)^c = \{\tau^* \succeq a\}.$$

Then

$$(\star) \qquad \sum_{a \prec \tilde{A}_n} \{\tau^* = a\} + \sum_{a \in \tilde{A}_n} \{\tau^* \succeq a\}$$

is, for each $n \in \mathbb{N}_0$, a partition of $\mathcal{X}$.
(1) For fixed $n \in \mathbb{N}_0$ the family

$$\mathcal{B}_n := \left\{ \begin{array}{l} C \in \mathcal{B} : C = \bigcup_{a \prec \tilde{A}_n} H_a \cup \bigcup_{a \in \tilde{A}_n} K_a \text{ where } H_a \in \mathcal{B}_a, \\ H_a \subset \{\tau^* = a\}, K_a \in \mathcal{B}_a, K_a \subset \{\tau^* \succeq a\} \end{array} \right\}$$

of the sets relevant for the first n sub-samples of τ^* form a σ-algebra over $\mathcal{X}$:

(i) For $H_a := \{\tau^* = a\}$ for $a \prec \tilde{A}_n$ and $K_a := \{\tau^* \succeq a\}$ for $a \in \tilde{A}_n$ one obtains according to $(\star)$
$$\mathcal{X} = \bigcup_{a \prec \tilde{A}_n} H_a \cup \bigcup_{a \in \tilde{A}_n} K_a.$$

(ii) For $C_i = \bigcup_{a \prec \tilde{A}_n} H_{a,i} \cup \bigcup_{a \in \tilde{A}_n} K_{a,i} \in \mathcal{B}_n, i \in \mathbb{N}$, follows
$$\bigcup_{i=1}^{\infty} C_i = \bigcup_{a \prec \tilde{A}_n} \bigcup_{i=1}^{\infty} H_{a,i} \cup \bigcup_{a \in \tilde{A}_n} \bigcup_{i=1}^{\infty} K_{a,i} \in \mathcal{B}_n.$$

(iii) For $C = \bigcup_{a \prec \tilde{A}_n} H_a \cup \bigcup_{a \in \tilde{A}_n} K_a$ one obtains according to (i)
$$C^c = \bigcup_{a \prec \tilde{A}_n} (\{\tau^* = a\} \backslash H_a) \cup \bigcup_{a \in \tilde{A}_n} (\{\tau^* \succeq a\} \backslash K_a) \in \mathcal{B}_n.$$

Moreover,

$$\mathcal{B}_n \subset \mathcal{B}_{n+1} \quad \forall n \in \mathbb{N}_0.$$

(2) The sequence $(V_n)_{n \in \mathbb{N}_0}$ where

$$V_n := \sum_{a \prec \tilde{A}_n} 1_{\{\tau^*=a\}} U_a + \sum_{a \in \tilde{A}_n} 1_{\{\tau^* \succeq a\}} \cdot U_a$$

forms a martingale with respect to $(\mathcal{B}_n)_{n \in \mathbb{N}_0}$: For each $n \in \mathbb{N}_0, a \preceq \tilde{A}_n$ and each random variable $X_a : (\mathcal{X}, \mathcal{B}_a) \to (\overline{\mathbb{R}}, \overline{\mathbb{B}})$ the function $1_{\{\tau^*=a\}} \cdot X_a$ is $\mathcal{B}_n$ measurable and for $h(a) = n$ also the function $1_{\{\tau^* \succeq a\}} \cdot X_a$. One obtains

$$\begin{aligned}
E(V_{n+1} \mid \mathcal{B}_n) &= E\left(\sum_{a \prec \tilde{A}_{n+1}} 1_{\{\tau^*=a\}} \cdot U_a \mid \mathcal{B}_n\right) + E\left(\sum_{a \in \tilde{A}_{n+1}} 1_{\{\tau^* \succeq a\}} \cdot U_a \mid \mathcal{B}_n\right)
\end{aligned}$$

(since the integrals over both conditional expectations are

bounded from above by $EX + \sup_{j \in \mathbb{N}_0} r_j$ and therefore $< \infty$ P-a.s.)

$$= \sum_{a \prec \tilde{A}_{n+1}} 1_{\{\tau^*=a\}} \cdot U_a + E\left(\sum_{a \in \tilde{A}_n} \sum_{j \in \tilde{M}_a} 1_{\{\tau^* \succeq aj\}} \cdot U_{aj} \mid \mathcal{B}_n\right)$$

$$= \sum_{a \prec \tilde{A}_n} 1_{\{\tau^*=a\}} \cdot U_a + \sum_{a \in \tilde{A}_n} 1_{\{\tau^*=a\}} \cdot U_a + \sum_{a \in \tilde{A}_n} \sum_{j \in \tilde{M}_a} 1_{\{\tau^* \succeq aj\}} \cdot U_a$$

since $E(1_{\{\tau^* \succeq aj\}} \cdot U_{aj} \mid \mathcal{B}_n) = 1_{\{\tau^* \succeq aj\}} \cdot E(U_{aj} \mid \mathcal{B}_a) = 1_{\{\tau^* \succeq aj\}} \cdot U_a$

according to the definition of τ^*

$$= V_n \quad P\text{-a.s.};$$

the integrability of V_n follows from $\mid EV_n \mid = \mid EV_0 \mid = \mid EU_{()} \mid < \infty$. Since

$$\sup_{n \in \mathbb{N}_0} E((-V_n)^-) = \sup_{n \in \mathbb{N}_0} E(V_n^+) \leq E(\sup_{a \in \mathcal{A}} Z_a^+) < \infty$$

the martingale convergence theorem (see e.g. [Ba], Satz 60.1) yields that $(-V_n)_{n \in \mathbb{N}_0}$ is P-a.s. convergent to an integrable random variable $(-V_\infty)$, i.e. $(V_n)_{n \in \mathbb{N}_0}$ converges P-a.s. to an integrable random variable V_∞.

(3) Assume that $\tau^* \notin \mathcal{A}$ with positive probability. Since by the construction of τ^*

$$P(\{\tau^* \in A \backslash \mathcal{A}\}) = 0$$

then $P(\{\tau^* \in A^*\}) > 0$. On the other hand

$$\begin{aligned}
P(\{\lim_{n \to \infty} V_n = -\infty\}) &= \\
&= P(\{\lim_{n \to \infty} (1_{\{\tau^* \prec \tilde{A}_n\}} \cdot Z_{\tau^*} + \sum_{a \in \tilde{A}_n} 1_{\{\tau^* \succeq a\}} \cdot U_a) = -\infty\}) \\
&\geq P(\{\lim_{n \to \infty} \sum_{a \in \tilde{A}_n} 1_{\{\tau^* \succeq a\}} \cdot U_a = -\infty\}) \\
&\geq P(\{\lim_{n \to \infty} \sum_{a \in \tilde{A}_n} 1_{\{\tau^* \succeq a\}} \cdot E(X + \sup_{j \geq n} r_j \mid \mathcal{B}_a) = -\infty\})
\end{aligned}$$

(since $U_a \leq E(X + \sup_{j \geq n} r_j \mid \mathcal{B}_a)$ P-a.s. $\forall a \in \tilde{A}_n$)

$$\geq P(\{\lim_{n\to\infty} 1_{\{\tau^* \succeq A_n\}} \sup_{j\geq n} r_j = -\infty\})$$

$$(\text{since } \lim_{n\to\infty} \sum_{a\in A_n} 1_{\{\tau^*\succeq a\}} E(X \mid \mathcal{B}_a) < \infty \ P\text{-a.s.})$$

$$= P(\{1_{\{\tau^*\in A^*\}} \cdot \limsup_{n\to\infty} r_n = -\infty\})$$

$$= P(\{\tau^* \in A^*\}) > 0 \ (\text{according to our assumption}).$$

But this contradicts the convergence of $(V_n)_{n\in\mathbb{N}_0}$ proven in (2); one obtains

$$P(\{\tau^* \in \mathcal{A}\}) = 1. \qquad \qquad \square$$

Of special importance for statistical applications is the case that the pay-off Z_a is, for $a = (a_1,\ldots,a_j) \in \mathcal{A}$, of the form

$$Z_a = Y_a - C(a)$$

where Y_a is a random part resulting from the terminal decisions and

$$C(a) = \sum_{i=1}^{j} c(a_i)$$

are the total costs of sampling (see (1.22)). For this case theorem (2.14) yields:

(2.15) Corollary

Let $Z_a = Y_a - C(a)$ s.t.

    (i) $E(\sup_{a\in\mathcal{A}} Y_a^+) < \infty$

and

    (ii) $C((a_1,\ldots,a_j)) = \sum_{i=1}^{j} c(a_i)$, $C(()) = 0$,

where $c : \mathrm{IN} \to (0; \infty)$ is a non-decreasing function with $\lim_{n\to\infty} c(n) = \infty$. Then τ^* is an optimal sampling plan.

Proof: (1) For $a \in \tilde{A}$ s.t. $\tilde{M}_a = \emptyset$ obviously

$$P(B(a\infty)) = 0;$$

for $a \in \tilde{A}$ s.t. $\tilde{M}_a \neq \emptyset$ follows

$$P(B(a\infty)) \leq P(\{(U_{ak} \mid \mathcal{B}_a) < \sup_{j\in\tilde{M}_a} E(U_{aj} \mid \mathcal{B}_a) \ \forall k \in \tilde{M}_a\})$$

$$\leq P(\{E(U_{ak} \mid \mathcal{B}_a) > E(U_{am} \mid \mathcal{B}_a) \text{ for infinitely many } k \in \tilde{M}_a\})$$

$$\text{where } m := \min \tilde{M}_a$$

$$= P(\{E(U_{ak} + C(ak) \mid \mathcal{B}_a) + c(m) - c(k) >$$

$$E(U_{am} + C(am) \mid \mathcal{B}_a) \text{ for infinitely many } k \in \tilde{M}_a\})$$

$$\leq P(\{c(m) - c(k) > E(U_{am} + C(am) \mid \mathcal{B}_a) -$$

$$- E(E(\sup_{b\in\mathcal{A}} Y_b^+ \mid \mathcal{B}_{ak}) \mid \mathcal{B}_a) \text{ for infinitely many } k \in \tilde{M}_a\})$$

$$= \quad P(\{E(U_{am} + C(am) - \sup_{b \in \mathcal{A}} Y_b^+ \mid \mathcal{B}_a) = -\infty\}),$$

$$\text{since } \lim_{n \to \infty} c(n) = -\infty$$

$$= \quad 0$$

$$\text{since } E(U_{am} + C(am) - \sup_{b \in \mathcal{A}} Y_b^+ \mid \mathcal{B}_a) \text{ is integrable by assumption,}$$

i.e. $P(B(a\infty)) = 0 \; \forall a \in \tilde{\mathcal{A}}$.

(2) For each $n \in \mathbb{N}_0$ and $a \in \mathcal{A}_n$ follows

$$Z_a = Y_a - C(a) \le X - C(a) \le X - n \cdot c(1) = X + r_n$$

where $X := \sup_{b \in \mathcal{A}} Y_b^+$ and $r_n := -n \cdot c(1)$. Since X is integrable by assumption and $\lim_{n \to \infty} r_n = -\infty$ according to $c(1) > 0$ the assertion now follows from theorem (2.14). $\qquad\square$

(2.16) Remark:

In the definition of τ^* the size of the next sub-sample is always chosen to be *minimal* under all $k \in \tilde{M}_a$ for which

$$(\star) \qquad\qquad E(U_{ak} \mid \mathcal{B}_a) = \sup_{i \in \tilde{M}_a} E(U_{ai} \mid \mathcal{B}_a)$$

(if the supremum is achieved). Corresponding optimality assertions hold if an "arbitrary" $k \in \tilde{M}_a$ with $(\star)$ is chosen (if the supremum is achieved) – τ^* is, in general, not the only optimal sampling plan.

§ 4 Optimal sampling plans for the Markov case

The previous results may be considerably simplified if the pay-off process $(Z_a)_{a \in \mathcal{A}}$ has a "Markov structure" (defined in the sequel) which has as a consequence that the future development of the process at each decision time depends only on the present "state" but not on the "way" this state was reached. We will show that one may, in these cases, restrict one's attention to sampling plans which take into consideration only the current "state" of the process.

(2.17) Definition

a) *Let* $(\mathcal{X}, \mathcal{B}, P)$, $(\mathcal{B}_a)_{a \in \mathcal{A}}$, $\mathcal{A}$ *and* $\tilde{\mathcal{A}}$ *be as in (2.1)/(2.8) and let* $(Y_a, \mathcal{B}_a)_{a \in \mathcal{A}}$ *be a stochastic process on* $(\mathcal{X}, \mathcal{B})$ *with values in* $(\mathcal{Y}, \mathcal{C})$.

(i) $(Y_a)_{a \in \mathcal{A}}$ *is called Markov process (with respect to* $(\mathcal{B}_a)_{a \in \mathcal{A}})$ *if*

$$P(Y_{ak} \in C \mid \mathcal{B}_a) = P(Y_{ak} \in C \mid Y_a) \quad P - a.s.$$

$\forall C \in \mathcal{C}, a \in \tilde{\mathcal{A}}, k \in \tilde{M}_a$.

(ii) *For* $\mathcal{A} = \tilde{\mathcal{A}} = A$ *a Markov process* $(Y_a)_{a \in A}$ *is called stationary if there exists, for each* $k \in M$, *a transition kernel*

$$Q_k : \mathcal{Y} \times \mathcal{C} \to [0; 1]$$

48

such that

$$Q_k(Y_a, C) = P(Y_{ak} \in C \mid \mathcal{B}_a) \quad P - a.s. \; \forall C \in \mathcal{C}, a \in A.$$

If there exists a $y \in \mathcal{Y}$ such that $Y_{()} \equiv y$ then y is called the initial state of the stationary Markov process $(Y_a)_{a \in A}$.

b) For a problem of optimal sequential sampling we have the Markov case if

$$Z_a = \eta_a \circ Y_a \quad \forall a \in \mathcal{A}$$

where $(Y_a)_{a \in \tilde{\mathcal{A}}}$ is a Markov process and η_a is, for each $a \in \mathcal{A}$, a measurable function

$$\eta_a : (\mathcal{Y}, \mathcal{C}) \to (\overline{\mathbb{R}^1}, \overline{\mathbb{B}^1}).$$

To illustrate these notions we consider example (2.3) once again:

(2.18) Example

Define in (2.3)

$$\mathcal{Y} := \mathbf{Z}, \; \mathcal{C} := \mathcal{P}(\mathbf{Z}), \; Y_a := \sum_{i=1}^{g(a)} X_i \text{ for } a \in \tilde{\mathcal{A}}$$

and

$$\eta_a := \begin{cases} y/h(a) & \text{for } a \succ () \\ \\ 0 & \text{for } a = () \end{cases}, \; a \in \tilde{\mathcal{A}}, y \in \mathbf{Z};$$

then

$$Z_a = \eta_a \circ Y_a$$

where $(Y_a)_{a \in \tilde{\mathcal{A}}}$ is a Markov process with respect to $(\mathcal{B}_a)_{a \in \tilde{\mathcal{A}}}$ since for $a \in \tilde{\mathcal{A}}, k \in \tilde{M}_a, C \in \mathcal{C}$

$$
\begin{aligned}
P(Y_{ak} \in C \mid \mathcal{B}_a) &= P(\sum_{i=1}^{g(ak)} X_i \in C \mid X_1, \ldots, X_{g(a)}) \\
&= P(\sum_{i=g(a)+1}^{g(ak)} X_i \in C - Y_a \mid X_1, \ldots, X_{g(a)}) \\
&= P(\sum_{i=g(a)+1}^{g(ak)} X_i \in C - y) \text{ where } y = Y_a(x), \\
&\quad \text{since the } X_i \text{ are independent} \\
&= P(Y_{ak} \in C \mid Y_a) \quad P - a.s.;
\end{aligned}
$$

therefore we are in the Markov case. Moreover, in the cases (2.3)(i) and (ii) the Markov process $(Y_a)_{a \in A}$ is stationary since

$$Q_k(y, C) := P(\sum_{i=1}^{k} X_i \in C - y)$$

yields a transition kernel with

$$Q_k(Y_a, C) = P(Y_{ak} \in C \mid \mathcal{B}_a) \quad P - a.s. \ \forall C \in \mathcal{C}, a \in A. \qquad \square$$

In the Markov case one obtains

$$P(Z_{ak} \in B \mid \mathcal{B}_a) = P(Z_{ak} \in B \mid Y_a) \quad \forall a \in \tilde{A}, k \in \tilde{M}_a, B \in I\!B,$$

i.e. the stochastic behaviour after time a depends only on the "state" Y_a at that time. Therefore it seems to be reasonable to consider sampling plans with a corresponding property:

(2.19) Definition
In the Markov case a randomized sampling plan β with $P^\beta(\mathcal{A}) = 1$ is called a Markovian sampling plan if the function $x \mapsto \beta_a(x, k)$ is $\sigma(Y_a)$-measurable for each $a \in \tilde{A}$ and $k \in \tilde{M}_a \cup \{0\}$.

To show that in the Markov case it is sufficient to consider only Markovian sampling plans we need the following auxiliary result:

(2.20) Lemma
$(Y_a)_{a \in \tilde{A}}$ is a Markov process with respect to $(\mathcal{B}_a)_{a \in \tilde{A}}$ iff for all $a \in \tilde{A}$, $k \in \tilde{M}_a$ and $B \in \mathcal{B}_a$
$$E(1_B \mid Y_{ak}) = E(E(1_B \mid Y_a) \mid Y_{ak}) \qquad P - a.s..$$
Proof: Let $a \in \tilde{A}, k \in \tilde{M}_a, B \in \mathcal{B}_a$ and $C \in \sigma(Y_{ak})$; then

$$\int_C E(1_B \mid Y_{ak}) \, dP = \int 1_C \cdot 1_B \, dP = \int_B E(1_C \mid \mathcal{B}_a) \, dP$$

and

$$\begin{aligned}
\int_C E(E(1_B \mid Y_a) \mid Y_{ak}) \, dP &= \int 1_C \cdot E(1_B \mid Y_a) \, dP \\
&= \int E(1_C E(1_B \mid Y_a) \mid Y_a) \, dP \\
&= \int E(1_C \mid Y_a) E(1_B \mid Y_a) \, dP = \int_B E(1_C \mid Y_a) \, dP.
\end{aligned}$$

Therefore

$$E(1_C \mid \mathcal{B}_a) = E(1_C \mid Y_a) \qquad P - a.s. \ \forall C \in \sigma(Y_{ak})$$

iff

$$E(1_B \mid Y_{ak}) = E(E(1_B \mid Y_a) \mid Y_{ak}) \qquad P - a.s. \ \forall B \in \mathcal{B}_a. \qquad \square$$

Algebraic induction yields:

(2.21) Corollary
$(Y_a)_{a \in \tilde{A}}$ is a Markov process with respect to $(\mathcal{B}_a)_{a \in \tilde{A}}$ iff for all $a \in \tilde{A}$, $k \in \tilde{M}_a$ and all (quasi-) integrable $f : (\mathcal{X}, \mathcal{B}_a) \to (\overline{I\!R^1}, \overline{I\!B_1})$

$$E(f \mid Y_{ak}) = E(E(f \mid Y_a) \mid Y_{ak}) \qquad P - a.s..$$

Using these characterizations we are now able to prove the following result:

(2.22) Theorem

In the Markov case there exists for each (randomized) sequential sampling plan β a randomized Markovian sampling plan $\tilde{\beta}$ such that

$$E(b_a^{\tilde{\beta}}(\cdot, k) \mid Y_a) = E(b_a^{\beta}(\cdot, k) \mid Y_a) \quad P - a.s. \qquad \forall a \in \tilde{A}, k \in \tilde{M}_a.$$

In particular

$$P^{\beta}(\{a\}) := \int b_a^{\beta}(x, 0) \, dP(x) = P^{\tilde{\beta}}(\{a\}) \qquad \forall a \in \mathcal{A}$$

and

$$E(Z_{\beta}) := E(\sum_{a \in \mathcal{A}} b_a^{\beta}(\cdot, 0) \, Z_a) = E(Z_{\tilde{\beta}}).$$

Proof: For a given sampling plan β and $a \in A \backslash \tilde{A}$ define

$$\tilde{\beta}_a(x, \cdot) := \varepsilon_0(\cdot);$$

for $a \in \tilde{A}$ and $k \in \tilde{M}_a \cup \{0\}$ let

$$\tilde{\beta}_a(x, k) := \begin{cases} f_a(x, k)/\overline{f_a}(x) & \text{if } \overline{f_a}(x) > 0 \\ \varepsilon_0(\{k\}) & \text{elsewhere} \end{cases}$$

where

$$f_a(\cdot, k) := E(b_a^{\beta}(\cdot, k) \mid Y_a)$$

and

$$\overline{f_a}(\cdot) := \sum_{k \in \tilde{M}_a \cup \{0\}} f_a(\cdot, k).$$

Assertion: $\tilde{\beta}$ has the desired properties.

Proof by induction on $h(a)$:

(i) $h(a) = 0$: For $k \in \tilde{M}_{()} \cup \{0\}$ from the definition of $\tilde{\beta}$ follows

$$b_{()}^{\tilde{\beta}}(\cdot, k) = \tilde{\beta}_{()}(\cdot, k) = E(b_a^{\beta}(\cdot, k) \mid Y_a),$$

i.e. the assertion.

(ii) Assume that the assertion holds true for all $a \in \tilde{A}$ s.t. $h(a) < n$, and let $a = (a_1, \ldots, a_n) \in \tilde{A}, k \in \tilde{M}_a \cup \{0\}$. Then

$$
\begin{aligned}
E(b_a^{\tilde{\beta}}(\cdot, k)|Y_a) \;&=\; E(\tilde{\beta}_a(\cdot, k)\, b_{(a_1,\ldots,a_{n-1})}^{\tilde{\beta}}(\cdot, a_n) \mid Y_a) \\[4pt]
&=\; \tilde{\beta}_a(\cdot, k)\, E(b_{(a_1,\ldots,a_{n-1})}^{\tilde{\beta}}(\cdot, a_n) \mid Y_a) \\[4pt]
&\stackrel{(2.21)}{=}\; \tilde{\beta}_a(\cdot, k)\, E(E(b_{(a_1,\ldots,a_{n-1})}^{\tilde{\beta}}(\cdot, a_n) \mid Y_{(a_1,\ldots,a_{n-1})}) \mid Y_a) \\[4pt]
&\stackrel{(assumption)}{=}\; \tilde{\beta}_a(\cdot, k)\, E(E(b_{(a_1,\ldots,a_{n-1})}^{\beta}(\cdot, a_n) \mid Y_{(a_1,\ldots,a_{n-1})}) \mid Y_a) \\[4pt]
&\stackrel{(2.21)}{=}\; \tilde{\beta}_a(\cdot, k)\, E(b_{(a_1,\ldots,a_{n-1})}^{\beta}(\cdot, a_n) \mid Y_a) \\[4pt]
&=\; \tilde{\beta}_a(\cdot, k)\overline{f}_a, \text{ since } b_{(a_1,\ldots,a_{n-1})}^{\beta}(\cdot, a_n) = \sum_{k \in \tilde{M}_a \cup \{0\}} b_a^{\beta}(\cdot, k) \\[4pt]
&=\; f_a(\cdot, k) = E(b_a^{\beta}(\cdot, k) \mid Y_a) \quad P - a.s.. \qquad \square
\end{aligned}
$$

This theorem shows that one may, when searching for optimal sequential sampling plans, restrict one's attention to (randomized) Markovian sampling plans: If there exists any optimal sampling plan then there exists also a (randomized) Markovian sampling plan with the same expected pay-off. Moreover, we will show that also the terms U_a and $E(U_{aj} \mid \mathcal{B}_a)$, $a \in \tilde{\mathcal{A}}, j \in \tilde{M}_a$, which turned out to be essential to the form of optimal sampling plans (see (2.5)/(2.8)), depend on $x \in \mathcal{X}$ by $Y_a(x)$ only.

For the case of a finite horizon, as treated in § 2, we obtain:

(2.23) Theorem

In the Markov case with finite A there exist $(\sigma(Y_a), \overline{\mathrm{IB}^1})$-measurable versions of U_a and of $E(U_{aj} \mid \mathcal{B}_a)$ $\forall a \in \tilde{\mathcal{A}}, j \in \tilde{M}_a$.

Proof by backward induction:

(i) For $a \in \tilde{\mathcal{A}}$ s.t. $\tilde{M}_a = \emptyset$ the assertion follows from $U_a = Z_a = \varphi_a \circ Y_a$.

(ii) Let $a \in \tilde{\mathcal{A}}$ s.t. $\tilde{M}_a \neq 0$ and assume that the assertion is fulfilled for all $b \in \tilde{\mathcal{A}}$ s.t. $b \succ a$. Then one obtains for each $j \in \tilde{M}_a$

$$E(U_{aj} \mid \mathcal{B}_a) = E(f(Y_{aj}) \mid \mathcal{B}_a) \text{ (by assumption for a measurable}$$
$$\text{function } f : (\mathcal{Y}, \mathcal{C}) \to (\overline{\mathrm{IR}^1}, \overline{\mathrm{IB}^1}))$$
$$= E(f(Y_{aj}) \mid Y_a) \quad P - a.s. \text{ (due to the Markov property)},$$

i.e. there exists a $\sigma(Y_a)$-measurable version of $E(U_{aj} \mid \mathcal{B}_a)$. Since by definition

$$U_a = \max\{\eta_a(Y_a), \max_{j \in \tilde{M}_a} E(U_{aj} \mid \mathcal{B}_a)\} \qquad P - a.s.$$

the assertion follows. $\qquad\qquad\square$

It may be conjectured that a corresponding assertion holds true also for the infinite horizon case. But as to the obvious idea to treat this case (similarly to the "time domain" $A = \mathrm{IN}$) by considering the limit

$$\max_{b \in \tilde{A}} g(b) \to \infty$$

the main tool, namely the triple limit theorem (see [C/R/S], pp. 81/104), is not available for sampling plans, as the following example shows:

(2.24) Example

Let $Y_i \equiv 0 \ \forall i \in \mathrm{IN}$, $X_i, i \in \mathrm{IN}$, independent s.t.

$$X_1 \equiv 1/2, \quad X_n = \begin{cases} n/(n-1) \\ \\ -n \end{cases} \text{ with probability } \begin{matrix} (n-1)/n \\ \\ 1/n \end{matrix} \ , n \geq 2,$$

and consider the problem of optimal sampling given by

$$M = \{1, 2\}, \ \mathcal{A} = \{\underbrace{(2, \ldots, 2, 1)}_{n\text{-times}} : n \in \mathrm{IN}_0\} \cup \{\underbrace{(2, \ldots, 2)}_{n\text{-times}} : n \in \mathrm{IN}\} \cup \{()\},$$

$$B_a = \begin{cases} \sigma(X_1, \ldots, X_{\bar{a}-1}, X_{\bar{a}}) & \text{if } a_{\bar{a}} = 1 \\ \sigma(X_1, \ldots, X_{\bar{a}-1}) & \text{if } a_{\bar{a}} = 2 \end{cases}$$

where $\bar{a} := n$ and $a_{\bar{a}} = a_n$ if $a = (a_1, \ldots, a_n) \in A$, and

$$Z_a := \begin{cases} X_{\bar{a}} & \text{if } a_{\bar{a}} = \begin{matrix} 1 \\ \\ 2 \end{matrix}, \qquad Z_{()} \equiv 0. \\ Y_{\bar{a}} \end{cases}$$

For $\alpha \leq -2$, $\beta \geq 2$ obviously (see [C/R/S], p. 77)

$$X_n(\alpha, \beta) = \begin{cases} X_n & n \leq -\alpha \\ & \text{if} \\ X_n' & n > -\alpha \end{cases}$$

where

$$X_n' = \begin{cases} n(n-1) & (n-1)/n \\ & \text{with probability} \\ \alpha & 1/n. \end{cases}$$

Then one obtains

$$v^N(\alpha, \beta) = 1 + \frac{\alpha}{N} > 1/2 \qquad \forall N > -2\alpha$$

and $v = 1/2$ (where $\tau^* = (1)$ is an optimal sampling plan) which leads to

$$V = \frac{1}{2} \neq 1 = \lim_{\beta \to \infty} \lim_{\alpha \to -\infty} \lim_{N \to \infty} v^N(\alpha, \beta). \qquad \qquad \square$$

Therefore we may merely conclude from (2.23):

(2.25) Corollary
Assume that in the Markov case

$$\lim_{m \to \infty} U_a^m = U_a \quad P - a.s. \quad \forall a \in \tilde{A}$$

holds for the problems defined by

$$A^m := \{a \in A : g(a) \leq m\}.$$

Then there exist $(\sigma(Y_a), \overline{\mathbb{B}^1})$-measurable versions of U_a $\forall a \in \tilde{A}$ and of $E(U_{aj}|\, B_a)$ $\forall j \in \check{M}_a, a \in \tilde{A}$.

A further simplification of the previous results can be obtained in the Markov case with a *stationary* process: Assume that $A = \tilde{A} = A$ and that $(Y_a^{(y)})_{a \in A}$ is, for each $y \in \mathcal{Y}$, a stationary Markov process with initial state y and transition kernels $Q_k, k \in M$ (independent of y; the underlying probability measure may depend on y). For $a \in A$ let

$$Z_a^{(y)} = \eta_a \circ Y_a^{(y)}$$

where we assume that for each $b \in A, ba := (b_1, \ldots, b_n, a_1, a_2, \ldots)$ and $y \in C$

$$E(\sup_{a \in A} \eta_{ba} \circ Y_a^{(y)}) < \infty \text{ and } E(\eta_{ba} \circ Y_a^{(y)}) > -\infty \quad \forall a \in A.$$

To consider problems of optimal sequential sampling for $\eta_{ba} \circ Y_a^{(y)}$ with an infinite horizon we denote for $N \in \mathbb{N}$

$$T_a^N := \{\tau \in T_a : g(\tau) \leq N\}.$$

According to our assumptions, $E(\eta_{b\tau} \circ Y_\tau^{(y)})$ exists for all $\tau \in T_a^N, b \in A$ and $y \in \mathcal{Y}$. Therefore we may define for $N \in \mathbb{N}, a \in A$ s.t. $g(a) \leq N, y \in \mathcal{Y}$ and $j \in M, j \leq N$,

$$U_a^{y,N} := \operatorname*{ess\,sup}_{\tau \in T_a^N} E(Z_\tau^{(y)} \mid \mathcal{B}_a)$$

and

$$v_{a,j,N}(y) := \sup\{E(\eta_{a\tau} \circ Y_\tau^{(y)}) : \tau \in T_{(j)}^N\}.$$

For these values one obtains

(2.26) Theorem

$$E(U_{aj}^{y,N} \mid \mathcal{B}_a) = v_{a,j,N-g(a)}(Y_a^{(y)}) \quad P-a.s.$$

for all $N \in \mathbb{N}, a \in A, j \in M$ s.t. $g(aj) \leq N$ and $y \in \mathcal{Y}$.

Proof: Let $M_a^N := \{j \in M : g(aj) \leq N\}$. For $N \in \mathbb{N}, a \in A$ s.t. $g(a) \leq N$ and $j \in M_a^N$ define

$$f_{N,a,j} : (\mathcal{Y}, \mathcal{C}) \to (\overline{\mathbb{R}^1}, \overline{\mathbb{B}^1})$$

by

$$f_{N,a,j}(s) := \begin{cases} \int\limits_{\mathcal{Y}} \eta_{aj}(y) \, Q_j(s, dy) & \text{if } M_{aj}^N = \emptyset \\ \int\limits_{\mathcal{Y}} \sup\{\eta_{aj}(y), f_{N,aj,k}(y) : k \in M_{aj}^N\} Q_j(s, dy) & \text{elsewhere.} \end{cases}$$

Assertion: $E(U_{aj}^{y,N} \mid \mathcal{B}_a) = f_{N,a,j}(Y_a^{(y)}) \quad P-a.s.$

The proof, performed by backward induction, is based on the fact that, due to the Bellman-equation, the $E(U_{aj}^{y,N} \mid \mathcal{B}_a)$ fulfill the same recursive relations as the $f_{N,a,j}$ do:

(i) Let $a \in A$ and $j \in M$ s.t. $M_{aj}^N = \emptyset$. Then P-a.s.

$$E(U_{aj}^{y,N} \mid \mathcal{B}_a) = E(\eta_{aj}(Y_{aj}^{(y)}) \mid \mathcal{B}_a) \quad (\text{see } (2.5))$$

$$= \int_{\mathcal{Y}} \eta_{aj}(y') \, Q_j(Y_a^{(y)}, dy')$$

$$\text{since } (Y_b^{(y)})_{b \in A} \text{ is a stationary Markov process}$$

$$= f_{N,a,j}(Y_a^{(y)}).$$

(ii) Let $a \in A, j \in M_a^N$ s.t. $M_{aj}^N \neq \emptyset$ and assume that

$$E(U_{ajk}^{y,N} \mid \mathcal{B}_{aj}) = f_{N,aj,k}(Y_{aj}^{(y)}) \quad P-a.s. \quad \forall k \in M_{aj}^N.$$

Then

$$\begin{aligned} E(U_{aj}^{y,N} \mid \mathcal{B}_a) &= E(\sup\{\eta_{aj}(Y_{aj}^{(y)}), E(U_{ajk}^{y,N} \mid \mathcal{B}_{aj}) : k \in M_{aj}^N\} \mid \mathcal{B}_a) \\ &\qquad \text{according to the Bellman-equation } (2.5) \\ &= E(\sup\{\eta_{aj}(Y_{aj}^{(y)}), f_{N,aj,k}(Y_{aj}^{(y)}) : k \in M_{aj}^N\} \mid \mathcal{B}_a) \end{aligned}$$

by assumption

$$= \int_{\mathcal{Y}} \sup\{\eta_{aj}(y'), f_{N,aj,k}(y') : \ k \in M_{aj}^N\} \, Q_j(Y_a^{(y)}, dy')$$

since $(Y_b^{(y)})_{b \in A}$ is a stationary Markov process

$$= f_{N,a,j}(Y_a^{(y)}) \qquad P - a.s.$$

Assertion: $v_{a,j,N-g(a)}(y) = f_{N,a,j}(y) \qquad \forall y \in \mathcal{Y}.$

To avoid misleading notation we define for $b \in A$ s.t. $g(b) \leq N - g(a)$

$$W_b^{y,N-g(a)} := \operatorname*{ess\,sup}_{\tau \in T_b^{N-g(a)}} E(\eta_{a\tau} \circ Y_\tau^{(y)} \mid \mathcal{B}_b).$$

Since

$$v_{a,j,N-g(a)}(y) = E(W_{(j)}^{y,N-g(a)}) \qquad \text{(see (2.5))}$$

it is sufficient to show that

$$E(W_{(j)}^{y,N-g(a)} \mid \mathcal{B}_{()}) = f_{N,a,j}(y) \quad P - a.s..$$

This follows from the fact that analogously to the proof of part (i)

$$E(W_{bj}^{y,N-g(a)} \mid \mathcal{B}_b) = f_{N,ab,j}(Y_b^{(y)}) \quad P - a.s.$$

for all $y \in \mathcal{Y}, j \in M$ and $b \in A$ s.t. $g(bj) \leq N - g(a)$. $\qquad \square$

Theorem (2.26) means that the optimum of what can be reached by intelligent sample plans for $(\eta_b \circ Y_b^{(y)})_{b \succeq a, g(b) \leq N}$ is the same as for $(\eta_{ab} \circ Y_b^{(z)})_{b \succeq (), g(b) \leq N - g(a)}$ where $z = Y_a^{(y)}(x)$. Out of all the information available at "time" $a \in A$ – represented by $\mathcal{B}_a$ – only $Y_a^{(y)}$ is essential for the decision to be made.

Again we conjecture that a corresponding assertion holds also for the infinite horizon case, i.e. that analogously to the theory of optimal stopping for

$$v_{a,j}^{(y)} := \sup_{\tau \in T_{(j)}} E(\eta_{a\tau}(Y_\tau^{(y)}))$$

and

$$U_a^{(y)} := \operatorname*{ess\,sup}_{\tau \in T_a} E(Z_\tau^{(y)} \mid \mathcal{B}_a)$$

follows

$$E(U_{aj}^{(y)} \mid \mathcal{B}_a) = v_{a,j}(Y_a^{(y)}) \qquad P - a.s..$$

But since a triple-limit-theorem is missing we can only conclude:

(2.27) Corollary

Assume that

$$\lim_{N \to \infty} U_a^{y,N} = U_a^y \quad P-a.s. \ and \ \lim_{N \to \infty} v_{a,j,N}(y) = v_{a,j}(y)$$

for all $a \in A$ and $y \in \mathcal{Y}$. Then

$$E(U_{aj}^{(y)} \mid \mathcal{B}_a) = v_{a,j}(Y_a^{(y)}) \quad P-a.s.$$

for all $a \in A, j \in M$ and $y \in \mathcal{Y}$.

Using the monotone convergence theorem this is an immediate consequence of (2.26).

A further simplification of the structure of optimal sequential sampling plans can be obtained if not only the process $(Y_a)_{a \in A}$ is stationary but also the functions η_a have a "stationarity" property:

(2.28) Lemma

Let the assumptions of (2.27) be fulfilled and

$$Z_a^{(y)} = \eta \circ Y_a^{(y)} \qquad \forall a \in A, y \in \mathcal{Y};$$

denote $v(y) := \sup_{\tau \in T} E(\eta(Y_\tau^{(y)}))$. Then $U_a^{(y)} = v(Y_a^{(y)}) \quad P-a.s.$ and

$$v_{a,j}(y) = \int_{\mathcal{Y}} v(s)\, Q_j(y, ds) =: v_j(y).$$

The Bellman-equations read

$$v(y) = \max\{\eta(y),\ \sup_{j \in M} v_j(y)\}.$$

Proof: Due to the form of Z_a one obtains

$$v_{a,j,N}(y) = v_{b,j,N}(y) \quad \forall a, b \in A,\ j \in M,\ N \in \mathrm{I\!N},\ y \in \mathcal{Y}$$

and therefore according to (2.27)

$$v_{a,j}(y) = v_{b,j}(y) =: \tilde{v}_j(y).$$

Hence (2.27) yields for the Bellman-equations

$$\begin{aligned} U_a^{(y)} &= \max\{\eta(Y_a^{(y)}), \sup_{j \in M} v_{a,j}(Y_a^{(y)})\} \\ &= \max\{\eta(Y_a^{(y)}), \sup_{j \in M} v_{(),j}(Y_a^{(y)})\} \\ &= v(Y_a^{(y)}). \end{aligned}$$

In particular it follows (since $Y_{()}^{(y)} = y$) for all $y \in \mathcal{Y}$

$$v(y) = \max\{\eta(y),\ \sup_{j \in M} \tilde{v}_j(y)\}.$$

56

Finally

$$\tilde{v}_j(y) \;=\; E(U^{(y)}_{(j)} \mid \mathcal{B}_{()}) = E(v(Y^{(y)}_{(j)}) \mid \mathcal{B}_{()})$$
$$=\; \int_{\mathcal{Y}} v(s)\, Q_j(y, ds) = v_j(y). \qquad \square$$

If, therefore, the plan τ^* from (2.9) fulfills $P(\{\tau^* \in A\}) = 1$ and the assumptions of (2.28) are satisfied then it follows for each $y \in \mathcal{Y}$ on the set $\{i < h \circ \tau^*\}$

$$U^{(y)}_{\tau^{*i}} = E(U^{(y)}_{\tau^{*i+1}} \mid \mathcal{B}_{\tau^{*i}}) = v_{\pi_{i+1}\circ\tau^*}(Y^{(y)}_{\tau^{*i}}) \quad P - a.s.$$

and

$$U^{(y)}_{\tau^*} = v(Y^{(y)}_{\tau^*}) \quad P - a.s. \text{ on } \{\tau^* \in A\}.$$

For $h \circ \tau^*$ one obtains P-a.s.

$$h \circ \tau^* = \inf\{i \in \mathrm{I\!N}_0 : Y^{(y)}_{\tau^{*i}} \in B\}$$

$(\inf \emptyset := \infty)$ where B denotes the $\mathcal{C}$-measurable set

$$B := \{\eta \geq v\},$$

which is independent of the initial state $y \in \mathcal{Y}$. This means that τ^* stops sampling as soon as the (present) state $Y^{(y)}_{\tau^{*i}}$ belongs to the "stopping set" B; if it belongs to the set

$$B_j := \{\eta < v, j = \inf\{i : v_i = \max_{k \in M} v_k\}\}$$

then τ^* continues sampling by a further sub-sample of size j.

(2.29) Example

Defining in example (2.3)(ii)

$$\mathcal{Y} \;:=\; \{(r, s) \in \mathrm{I\!N}_0 \times \mathbf{Z} : s \leq m \cdot r\},\ \mathcal{C} = \mathcal{P}(\mathcal{Y}),$$
$$Y_a \;:=\; (h(a), \sum_{i=1}^{g(a)} X_i) \text{ for } a \in A$$

and

$$\eta(r, s) := \begin{cases} s/r & r > 0 \\ & if \\ 0 & r = 0 \end{cases}$$

one obtains that $Z_a = \eta \circ Y_a$ and that $(Y_a)_{a \in A}$ is a Markov process where

$$Q_j((r, s), C) := P((r + 1, s + \sum_{i=1}^{j} X_i) \in C), \quad j \in M, (r, s) \in \mathcal{Y}, C \in \mathcal{C}$$

yields a transition kernel such that

$$Q_j(Y_a, C) \;=\; P((Y_a + 1, \sum_{i=g(a)+1}^{g(a)+j} X_i) \in C \mid Y_a)$$
$$=\; P(Y_{aj} \in C \mid Y_a) = P(Y_{aj} \in C \mid \mathcal{B}_a) \quad P - a.s..$$

This means that one has a stationary Markov process $(Y_a)_{a\in A}$ with transition kernels Q_j, initial state $(0,0)$ and a "stationary" function η.

Similarly, $(Y_a^{(r,s)})_{a\in A}$ where

$$Y_a^{(r,s)} := (r,s) + Y_a \ \text{ for } (r,s) \in \mathcal{Y}$$

forms a stationary Markov process with transition kernels Q_j and initial state (r,s). For this process

$$U_a^{(r,s)} = \lim_{N\to\infty} U_a^{(r,s),N} \quad P-a.s. \quad \forall a \in A :$$

Defining for $\tau \in T_a$ and $N \in \mathrm{IN}$ s.t. $N > g(a)$

$$\tau^N := \begin{cases} \tau & \text{if } g(\tau) \leq N \\ \max\{\tau^i : g(\tau^i) \leq N, i \in \mathrm{IN}_0\} & \text{elsewhere} \end{cases}$$

one has $\tau^N \in T_a^N$; for $(r,s) \in \mathcal{Y}$ one obtains

$$\begin{aligned}
E(Z_\tau^{(r,s)} \mid \mathcal{B}_a) &= E(Z_{\tau^N}^{(r,s)} + 1_{\{g(\tau)>N\}}(Z_\tau^{(r,s)} - Z_{\tau^N}^{(r,s)}) \mid \mathcal{B}_a) \\
&\leq U_a^{(r,s),N} + 2m \cdot P(\{g(\tau) > N\} \mid \mathcal{B}_a) \quad P-a.s.
\end{aligned}$$

since $\mid Z_\tau^{(r,s)} - Z_{\tau^N}^{(r,s)} \mid \leq 2m$ and, due to $P(\{g(\tau) > N\}) \stackrel{N\to\infty}{\to} 0$,

$$E(Z_\tau^{(r,s)} \mid \mathcal{B}_a) \leq \lim_{N\to\infty} U_a^{(r,s),N} \quad P-a.s..$$

The inversed inequality is obviously fulfilled; hence

$$U_a^{(r,s)} = \lim_{N\to\infty} U_a^{(r,s),N} \quad P-a.s..$$

Analogously it follows for all $(r,s) \in \mathcal{Y}$ and $a \in A, j \in M$ that

$$v_{a,j}((r,s)) = \lim_{N\to\infty} v_{a,j,N}((r,s)).$$

Using the result of example $(2.13)a)$ one obtains that an optimal sequential sampling plan may be described by the stopping set $B := \{\eta \geq v\}$ and by the "continuation set" B^c where sampling is continued with a sub-sample of size m. ⊓

III. Sequentially planned tests; sequentially planned probability ratio tests

§ 1 Notation

An important type of statistical decision problems is characterized by the property that only two different (terminal) decisions concerning the parameter ϑ are possible, i.e. that two disjoint subsets H_1 and H_2 of Θ are selected[25] and that

$$D = \{d_1, d_2\}, \quad \mathcal{D} = \mathcal{P}(D)$$

where d_i is interpreted as "H_i holds true" ($\vartheta \in H_i$). In this case one speaks of a (statistical) *testing problem*; the sets H_i are called *hypotheses*. The examples (1.7)/(1.14)a) and (1.8)/(1.14)b) are concerned with problems of this type.

Obviously a randomized terminal decision function $\varphi_a, a \in A$, may be characterized by the probabilities

$$\varphi_a(x, d_2)$$

of choosing, for an observed $x \in \mathcal{X}$, the decision d_2. Simplifying the notations we use $\varphi_a(x)$ instead of $\varphi_a(x, d_2)$; the $\mathcal{B}_a$-measurable functions

$$\varphi_a : \mathcal{X} \to [0; 1]$$

defined in this way are called *tests* (of the hypotheses H_i). Analogously, a sequentially planned statistical decision procedure (τ, φ) is called *sequentially planned test*.

For the "quality" of a sequentially planned test the probabilities of erroneous decisions will be important as well as the costs of sampling:

(3.1) Definition
    Let (τ, φ) be a sequentially planned test.
    (i) The function $PF : \Theta \to [0; 1]$ defined by $\vartheta \mapsto E_\vartheta \varphi_\tau$ is called <u>p</u>ower <u>f</u>unction of (τ, φ). $OC := 1\text{-}PF$ is called <u>o</u>perating <u>c</u>haracteristic (OC-function) of (τ, φ).
    (ii) The function $ASC : \Theta \to [0; \infty]$ defined by $\vartheta \mapsto E_\vartheta\, C(\vartheta, \tau)$ is called the <u>a</u>verage <u>s</u>ampling <u>c</u>osts function (ASC-function).

For a sequentially planned test (τ, φ) and $\vartheta \in H_1$

$$\alpha(\vartheta, (\tau, \varphi)) := PF(\vartheta) = E_\vartheta\, \varphi_\tau$$

gives the probability of the first kind of error at point ϑ; analogously, for $\vartheta \in H_2$

$$\beta(\vartheta, (\tau, \varphi)) := OC(\vartheta) = E_\vartheta(1 - \varphi_\tau)$$

is the probability of the second kind of error at point ϑ.

Looking for reasonable/optimal sequentially planned tests one will try to keep the probabilities of the first kind of error "small" as well as the probabilities of the second kind of

---

[25]It will always be assumed that $P_{\vartheta_1} \neq P_{\vartheta_2}$ for $\vartheta_1 \neq \vartheta_2$ i.e. no superfluous parameter values are considered.

error and, moreover, also the ASC-function. But these contrary aims can simultaneously be achieved only in trivial exceptions. Therefore one has to find compromises – some reasonable kinds of compromises (Bayes-tests, tests at given error-level, tests with given maximal costs...) will be discussed in subsequent chapters.

From the Neyman-Pearson lemma it is well-known that likelihood ratios play an important role for the discrimination of statistical hypotheses. Since orthogonal parts cause no problems for this discrimination (it is trivial to choose the correct decision) but need, in many cases, special formulations, we make (if not explicitly excluded) the general assumption (see also [Ir], p.12)

(A) For all $\vartheta, \eta \in \Theta$ and all $a \in A$ the probability measures $P_\vartheta \mid \mathcal{B}_a$ and $P_\eta \mid \mathcal{B}_a$ are equivalent, i.e. for $B \in \mathcal{B}_a$ one gets $P_\vartheta(B) = 0$ iff $P_\eta(B) = 0$.

This assumption by no means implies the equivalence of $P_\vartheta \mid \mathcal{B}$ and $P_\vartheta \mid \mathcal{B}$; in fact, for problems with unbounded total sample sizes different probability measures P_ϑ, P_η will, in general, be orthogonal.

Under assumption (A) there exists, for each $a \in A$, a Radon-Nikodym derivative

$$dP_\vartheta \mid \mathcal{B}_a \; / \; dP_\eta \mid \mathcal{B}_a.$$

(3.2) Definition
Each version $q_a(\vartheta, \eta)$ of the Radon-Nikodym derivative

$$dP_\vartheta \mid \mathcal{B}_a \; / \; dP_\eta \mid \mathcal{B}_a, \quad \vartheta, \eta \in \Theta, \; a \in A$$

is called likelihood ratio of the sample sequence a.

Since $0 < q_a(\vartheta, \eta) < \infty$ a.s., also the logarithmic likelihood ratio exists:

$$-\infty < \log q_a(\vartheta, \eta) < \infty.$$

If no observational data are available without sampling, i.e. if $\mathcal{B}_{()} = \{\emptyset, \mathcal{X}\}$, then $q_{()}(\vartheta, \eta) \equiv 1 \quad \forall \vartheta, \eta \in \Theta$.

From the likelihood ratios $q_a(\vartheta, \eta), a \in A$, one obtains in a simple way, for a sequential sampling plan τ, likelihood ratios for the σ-algebra

$$\mathcal{B}_\tau := \{B \in \mathcal{B} : \; B \cap \{\tau = a\} \in \mathcal{B}_a \quad \forall a \in A\}$$

of the τ-history, too:

(3.3) Lemma *(see [Ir], 1.2.4)*
Let τ be a sequential sampling plan s.t. $P_\vartheta(\tau \in A) = 1 \; \forall \vartheta \in \Theta$ and $\vartheta, \eta \in \Theta$. Then $P_\vartheta \mid \mathcal{B}_\tau$ and $P_\eta \mid \mathcal{B}_\tau$ are equivalent and

$$dP_\vartheta \mid \mathcal{B}_\tau \; / \; dP_\eta \mid \mathcal{B}_\tau = q_\tau(\vartheta, \eta),$$

i.e.

$$P_\vartheta(B) = \int_B q_\tau(\vartheta, \eta) \, dP_\eta \quad \forall B \in \mathcal{B}_\tau.$$

60

Proof: By definition, $B \cap \{\tau = a\} \in \mathcal{B}_a$ for $B \in \mathcal{B}_\tau$; therefore

$$\begin{aligned}
P_\vartheta(B) &= \sum_{a \in A} P_\vartheta(B \cap \{\tau = a\}) \text{ since } P_\vartheta(\tau \in A) = 1 \\
&= \sum_{a \in A} \int_{B \cap \{\tau = a\}} q_a(\vartheta, \eta)\, dP_\eta = \int_B q_\tau(\vartheta, \eta)\, dP_\eta. \qquad \square
\end{aligned}$$

§ 2   The iid case

In the field of mathematical stochastics the case of independent repetition of experiments, i.e. the observation of independent identically distributed random variables, is of dominant importance. Therefore the case

$$\mathcal{B}_a = \sigma(X_1, \ldots, X_{g(a)}), \ a \in A \ \text{ (in particular } \mathcal{B}_{()} = \{\emptyset, \mathcal{X}\})$$

where $(X_i)_{i \in \mathbb{N}}$ is a sequence of independent identically distributed random variables is of special interest – we call it the *iid case*.

If τ is a sequential sampling plan, then

$$N := g \circ \tau$$

(where $g(a) := \infty$ for $a \in A^*$) yields a stopping rule with respect to $(\mathcal{B}_{(n)})_{n \in \mathbb{N}}$ since

$$\{g \circ \tau = n\} = \cup_{a \in A}\{\tau = a, g(a) = n\} \in \sigma(X_1, \ldots, X_n) = \mathcal{B}_{(n)}, \ n \in \mathbb{N}.$$

Using the notation $S_n := \sum_{i=1}^{n} X_i$ one therefore obtains from Wald's equation (see e.g. [Ir], 1.1.3)

(3.4) Remark *("Wald's equation")*
   In the iid case one obtains for each sequential sampling plan τ:

   a) $E_\vartheta(S_{g \circ \tau}) = E_\vartheta(X_1)\, E_\vartheta(g \circ \tau)$, *if $E_\vartheta(X_1)$ exists and $E_\vartheta(g \circ \tau) < \infty$,*

   b) $E_\vartheta(S^2_{g \circ \tau}) = E_\vartheta(X_1^2)\, E_\vartheta(g \circ \tau)$, *if $E_\vartheta(X_1) = 0$, $E_\vartheta(X_1^2) < \infty$ and $E_\vartheta(g \circ \tau) < \infty$.*

If the probability measures $P_\vartheta^{X_1}$ are, moreover, dominated by a σ-finite measure μ, i.e. if there exist densities

$$f_\vartheta := dP_\vartheta^{X_1}/d\mu,$$

then we speak of the *dominated iid case*. Using the notation

$$f(\vartheta, \eta) := dP_\vartheta^{X_1}/dP_\eta^{X_1} = f_\vartheta/f_\eta, \qquad \vartheta, \eta \in \Theta,$$

one then obtains (see [Ir], p. 17) for the likelihood ratio

$$q_a(\vartheta, \eta) = \prod_{i=1}^{g(a)} f(\vartheta, \eta)(X_i), \quad a \in A,$$

and for the logarithmic likelihood ratios

$$\log q_a(\vartheta, \eta) = \sum_{i=1}^{g(a)} \log f(\vartheta, \eta)(X_i),$$

where $(\log f(\vartheta, \eta)(X_i))_{i \in \mathbb{N}}$ forms a sequence of independent identically distributed random variables. The *Kullback-Leibler-information*

$$I(\vartheta, \eta) := E_\vartheta(\log f(\vartheta, \eta)(X_1))$$

exists (in the wider sense, i.e. $+\infty$ is possible) and it follows $I(\vartheta, \eta) > 0$.

§ 3 Sequentially planned probability ratio tests

For the development of sequential analysis the sequential probability ratio tests (SPRT's; see (1.8)/(1.14)b)), defined and intensively investigated already by A. Wald, played a dominant role. For an illustration of our previous notion we therefore consider sequentially planned tests which turn out to be, in particular for non-linear cost functions, reasonable generalizations of these SPRT's.

For the dominated iid case, i.e.

$$(\mathcal{X}, \mathcal{B}) = (\mathrm{IR}^{\mathbb{N}}, I\!B^{\mathbb{N}}), \quad P_\vartheta = (Q_\vartheta)^{\mathbb{N}},$$

let $f_\vartheta = dQ_\vartheta/d\mu$ be densities with respect to a σ-finite dominating measure μ, let

$$M = \mathbb{N} \quad \text{and} \quad \mathcal{A} = A = \cup_{j=1}^{\infty} \mathbb{N}^j \cup \{()\}$$

(i.e. in particular infinite horizon) and

$$\mathcal{B}_a = \pi_{\{1,\ldots,g(a)\}}^{-1}(I\!B^{g(a)}), \qquad a \in A.$$

Moreover, let $\vartheta_1, \vartheta_2 \in \Theta$ (s.t. $Q_{\vartheta_1} \neq Q_{\vartheta_2}$), H_1 and H_2 be disjoint sub-sets of Θ s.t. $\vartheta_i \in H_i, i = 1, 2$, and

$$D = \{d_1, d_2\}, \quad \mathcal{D} = \mathcal{P}(D).$$

This means that one has a hypotheses testing problem for independent repetitions of experiments where two special distributions are selected. Let now $k_1 < 1 < k_2$ be given constants and $\hat{t} : \mathrm{IR} \to \mathrm{IN}_0$ such that $\{\hat{t} = n\} \in I\!B \forall n \in \mathrm{IN}_0$ and

$$\hat{t}(x) = \left\{ \begin{array}{ll} = 0 & x \notin (k_1; k_2) \\ & \text{for} \\ > 0 & x \in (k_1; k_2). \end{array} \right.$$

(3.5) Definition

In the previous situation define

$$\tau_1 := \hat{t}(1), \quad \tau_{n+1} := \hat{t}(q_{(\tau_1,\ldots,\tau_n)}(\vartheta_2, \vartheta_1)), \quad n \in \mathrm{IN}.$$

The decision procedure (τ, φ) defined by

$$\tau := (\tau_1, \ldots, \tau_{\min\{j : \tau_j = 0\} - 1})$$

62

and

$$\varphi_a := \begin{cases} 0 & \\ & \text{if } q_a(\vartheta_2, \vartheta_1) \quad \begin{matrix} \le \\ > \end{matrix} \quad 1 \\ 1 & \end{cases}$$

is called sequentially planned probability ratio test (SPPRT).

The special case $\hat{t}(x) = 1 \quad \forall x \in (k_1, k_2)$ just yields the (purely) sequential probability ratio tests (SPRT's). But also for more general functions $\hat{t}$ the procedure is quite similar to the SPRT's (see (1.8)): Applying an SPPRT (τ, φ)

- one makes observations until $q_{(\tau_1,\ldots,\tau_n)}(\vartheta_2, \vartheta_1)$ leaves the interval (k_1, k_2); within the interval (k_1, k_2) the next (sub-) sample size is determined according to $\hat{t}$,

- one decides in favour of H_1 if $q_\tau(\vartheta_2, \vartheta_1) \le k_1$ and in favour of H_2 if $q_\tau(\vartheta_2, \vartheta_1) \ge k_2$.

A graphical illustration of such a procedure is given in figure 3.1 (by means of three different sample paths)

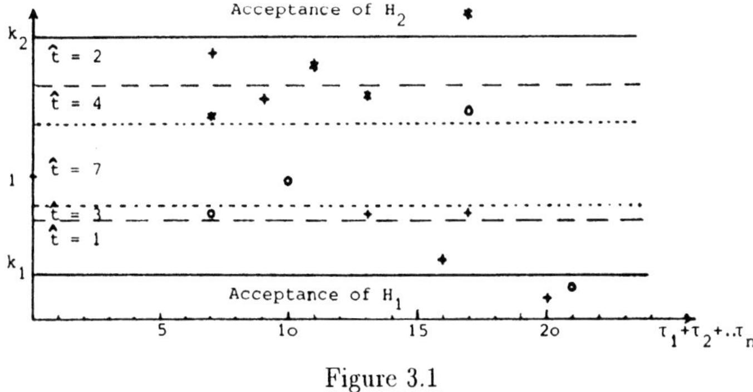

Figure 3.1

(the sample path marked by o leads on the basis of 4 sub-samples and a total sample size of 21, to the acceptance of H_1, that marked by $\star$ after 4 sub-samples with a total number of 17 observations to the acceptance of H_2, and that marked by $+$ on the basis of 5 sub-samples with 20 observations to the acceptance of H_1)

In order to prove that definition (3.5) makes sense, it remains to be shown that τ is really a sequential sampling plan which leads to a final decision – since the $\mathcal{B}_a$-measurability conditions are fulfilled by definition one needs:

(3.6) Remark
For the mapping τ defined in (3.5) holds

$$P_\vartheta(\tau \in A) = 1 \qquad \forall \vartheta \in \Theta.$$

To avoid superfluous repetitions we postpone the proof until theorem (3.9).

In (1.9) we gave inequalities and approximations for the error probabilities at the parameter values ϑ_i – the only important property was that the likelihood ratio lies outside the interval $(k_1; k_2)$ when sampling is finished. These results can immediately be generalized to SPPRT's:

(3.7) Lemma

Let (τ, φ) be a sequentially planned probability ratio test (as in (3.5)).
Then
a) $\alpha(\vartheta_1, (\tau, \varphi)) \leq (1 - \beta(\vartheta_2, (\tau, \varphi)))/k_2$,
* $\beta(\vartheta_2, (\tau, \varphi)) \leq (1 - \alpha(\vartheta_1, (\tau, \varphi))) \, k_1$.*
b) For $k_1 = \beta, k_2 = 1/\alpha$ where $0 < \alpha, \beta < 1$

$$\alpha(\vartheta_1, (\tau, \varphi)) \leq \alpha, \qquad \beta(\vartheta_2, (\tau, \varphi)) \leq \beta.$$

c) For $k_1 = \beta/(1 - \alpha), k_2 = (1 - \beta)/\alpha$ where $0 < \alpha, \beta < 1, \ \alpha + \beta < 1$,

$$\alpha(\vartheta_1, (\tau, \varphi)) + \beta(\vartheta_2, (\tau, \varphi)) \leq \alpha + \beta.$$

To *prove* these inequalities one only has to mention that according to (3.3)

$$\alpha(\vartheta_1, (\tau, \varphi)) = E_{\vartheta_1}\varphi_\tau = E_{\vartheta_2}(q_\tau(\vartheta_1, \vartheta_2)\,\varphi_\tau)$$

$$= E_{\vartheta_2}(q_\tau^{-1}(\vartheta_2, \vartheta_1)\,\varphi_\tau) \leq \frac{1}{k_2}E_{\vartheta_2}\varphi_\tau = (1 - \beta(\vartheta_2(\tau, \varphi)))/k_2$$

and analogously

$$\beta(\vartheta_2, (\tau, \varphi)) = E_{\vartheta_1}(q_\tau(\vartheta_2, \vartheta_1)(1 - \varphi_\tau)) \leq (1 - \alpha(\vartheta_1, (\tau, \varphi))) \, k_1;$$

parts b) and c) are obtained as in (1.9). $\qquad\qquad\qquad\square$

An obvious consequence of (3.7)a) is

$$\alpha(\vartheta_1, (\tau, \varphi)) \leq 1 - \beta(\vartheta_2, (\tau, \varphi)),$$

i.e. SPPRT's are *unbiased* for the simple hypotheses $\{\vartheta_1\}, \{\vartheta_2\}$.

(1.9)/(3.7) give hints how to choose, for given levels α and β, the stopping bounds k_i in a reasonable way. But since these inequalities are based on neglecting the excess (overshoot over the bounds k_i) and this excess will, due to a possible accumulation of several overshoots, in general be larger for SPPRT's than for SPRT's, the "approximations" of (3.7) will be more crude than those of (1.9). For this reason and since one needs indications for the error probabilities at parameter values $\vartheta \neq \vartheta_i$, algorithms for (numerically) computing the OC-function are of special interest (comp. § 4).

Besides the OC-function, the ASC-function is important for judging the "quality" of sequentially planned tests. But the next example (with the especially simple cost function $c(\vartheta, n) = c \cdot n$) shows that one needs additional assumptions to ensure that the expectations $E_\vartheta \, C(\vartheta, \tau)$ are finite:

(3.8) Example

Let $P_1^{X_1}(\{n\}) = 2^{-n}$, $P_2^{X_1}(\{n\}) = 3(2^{-n-1} - 2^{-2n-1})$, $n \in \mathrm{IN}$; then

$$P_2^{X_1}(\{n\})/P_1^{X_1}(\{n\}) = \frac{3}{2}(1 - \frac{1}{2^n}).$$

Defining for $k_1 = \frac{1}{2}, k_2 = 2$

$$\hat{t}(1) = 1, \ \hat{t}(\frac{3}{2}(1 - \frac{1}{2^n})) = 2^n$$

(and arbitrary > 0 elsewhere on (k_1, k_2)), one obtains for $c(\vartheta, n) = c \cdot n$

$$E_1 \, C(1, \tau) = c \, E_1(g \circ \tau) \geq c \, E_1(\tau_1 + \tau_2) \geq c(1 + \sum_{n=1}^{\infty} \frac{1}{2^n} \cdot 2^n) = \infty;$$

in particular there does not exist any moment of $N = g \circ \tau$. $\qquad\qquad\square$

But assuming that the function $\hat{t}$ is bounded, i.e. that there exists an upper bound for the size of sub-samples (comp. [Ha]), one can – similarly as for Wald's SPRT's – even prove the exponential boundedness of $N = g \circ \tau$:

(3.9) Theorem

Let (τ, φ) an SPPRT with $\hat{t} \leq m \in \mathrm{IN}$. Then for each $\vartheta \in \Theta$
a) there exist $R > 0$ and $\rho \in (0, 1)$ such that

$$P_\vartheta(g \circ \tau \geq j) \leq R\rho^j, \quad j \in \mathrm{IN};$$

b) $E_\vartheta(g \circ \tau)^n < \infty \quad \forall n \in \mathrm{IN}$.

Proof (see also [St2]): From (A) follows

$$P_\vartheta(f(\vartheta_2, \vartheta_1)(X_1) = 0) = 0 = P_\vartheta(f(\vartheta_2, \vartheta_1)(X_1) = \infty),$$

hence $\log q_a(\vartheta_2, \vartheta_1)$ always exists. Moreover $P_\vartheta(f(\vartheta_2, \vartheta_1)(X_1) \neq 1) > 0$ (according to footnote 25), i.e. there exist $\varepsilon > 0$ such that $P_\vartheta(\log f(\vartheta_2, \vartheta_1)(X_1) > \varepsilon) > 0$ or $P_\vartheta(\log f(\vartheta_2, \vartheta_1)(X_1) < -\varepsilon) > 0$. To prove a) choose $n_0 \in \mathrm{IN}$ such that

$$n_0 > 2m \ \text{ and } \ (n_0 - 2m)\epsilon > \log k_2 - \log k_1.$$

For $k \in \mathrm{IN}$ define

$$A_k := \left\{ \begin{array}{l} | \sum_{i=n_1}^{n_2} \log f(\vartheta_2, \vartheta_1)(X_i) | > \log k_2 - \log k_1 \\[2mm] \forall n_1, n_2 : \begin{array}{l} (k-1)n_0 < n_1 \leq (k-1)n_0 + m + 1 \\ k \, n_0 - m < n_2 \leq k \, n_0 \end{array} \end{array} \right\}.$$

The events $A_k, k \in \mathrm{IN}$, are independent since A_k is determined by $X_{(k-1)n_0+1}, \ldots, X_{kn_0}$, and $P_\vartheta(A_k) = P_\vartheta(A_1) \, \forall k \in \mathrm{IN}$.

(i) $\qquad\qquad \delta := P_\vartheta(A_1) > 0.$

This follows from

$$P_\vartheta(\log f(\vartheta_2,\vartheta_1)(X_1) > \varepsilon)^{n_0} = P_\vartheta(\log f(\vartheta_2,\vartheta_1)(X_i) > \varepsilon,\ 1 \leq i \leq n_0),$$

$$\text{since the } X_i \text{ are independent}$$

$$\leq P_\vartheta\Big(\sum_{i=n_1}^{n_2} \log f(\vartheta_2,\vartheta_1)(X_i) > (n_2-n_1+1)\,\varepsilon\ \forall n_1,n_2 : \begin{array}{c} 0 < n_1 \leq m+1 \\ n_0 - m < n_2 \leq n_0 \end{array} \Big)$$

$$\leq P_\vartheta(A_1)$$

since

$$(n_2 - n_1 + 1) \geq (n_0 - m) + 1 - (m+1) = n_0 - 2m$$

and therefore $(n_2 - n_1 + 1)\,\varepsilon > \log k_2 - \log k_1$.

Analogously one obtains

$$P_\vartheta(\log f(\vartheta_2,\vartheta_1)(X_1) < -\varepsilon)^{n_0} \leq P_\vartheta(A_1);$$

altogether

$$P_\vartheta(A_1) \geq \max\{P_\vartheta(\log f(\vartheta_2,\vartheta_1)(X_1) > \varepsilon)^{n_0},$$

$$P_\vartheta(\log f(\vartheta_2,\vartheta_1)(X_1) < -\varepsilon)^{n_0})\} > 0.$$

(ii) $$\qquad \bigcup_{j \leq k} A_j \subset \{g \circ \tau \leq k\,n_0\}.$$

For a proof let $j \leq k$ and $x \in A_j$. If $g \circ \tau(x) \leq j\,n_0 - m$ then also $x \in \{g \circ \tau \leq k\,n_0\}$. Consider therefore the case $g \circ \tau(x) > j\,n_0 - m$. From $\tau_i(x) \leq m\ \ \forall i$ and $n_0 > 2m$ follows the existence of $m_1, m_2 \in$ IN, $m_1 < m_2$, such that

$$(j-1)\,n_0 < \tau_1(x) + \ldots + \tau_{m_1}(x) \leq (j-1)\,n_0 + m,$$

$$j\,n_0 - m < \tau_1(x) + \ldots + \tau_{m_2}(x) \leq j\,n_0;$$

in particular

$$\sum_{i=1}^{\tau_1(x)+\ldots+\tau_{m_1}(x)} \log f(\vartheta_2,\vartheta_1)(x) \in (\log k_1; \log k_2).$$

On the other hand, $x \in A_j$ yields

$$\left| \sum_{i=\tau_1(x)+\ldots+\tau_{m_1}(x)+1}^{\tau_1(x)+\ldots+\tau_{m_2}(x)} \log f(\vartheta_2,\vartheta_1)(x) \right| > \log k_2 - \log k_1$$

and therefore

$$\sum_{i=1}^{\tau_1(x)+\ldots+\tau_{m_2}(x)} \log f(\vartheta_2,\vartheta_1)(x) \notin (\log k_1; \log k_2).$$

hence

$$g \circ \tau(x) \leq \tau_1(x) + \ldots + \tau_{m_2}(x) \leq j\,n_0 \leq k\,n_0.$$

(iii) $\qquad$ From (ii) follows for $k \in$ IN

$$P_\vartheta(g \circ \tau > n_0\,k) \leq P_\vartheta\Big(\bigcap_{j \leq k} A_j^c \Big)$$

$$= (P_\vartheta(A_1^c))^k, \text{ since the } A_k \text{ are independent and } P_\vartheta(A_k) = P_\vartheta(A_1) \; \forall k \in \text{IN}$$
$$\overset{(i)}{\leq} (1-\delta)^k.$$

If $\delta = 1$ one obtains $g \circ \tau \leq n_0 \, k \quad P_\vartheta - a.s.$ and therefore immediately the assertion. Elsewhere choose for $j \in \text{IN}$ a $k \in \text{IN}$ such that

$$(k-1)\, n_0 < j \leq k\, n_0;$$

then

$$\begin{aligned}
P_\vartheta(g \circ \tau \geq j) &\leq P_\vartheta(g \circ \tau > (k-1)\, n_0) \leq (1-\delta)^{k-1} \\
&= \frac{1}{1-\delta}((1-\delta)^{1/n_0})^{k n_0} \leq \frac{1}{1-\delta}((1-\delta)^{1/n_0})^j.
\end{aligned}$$

Defining $R = 1/(1-\delta)$ and $\rho = (1-\delta)^{1/n_0}$ one therefore obtains assertion a), i.e. the *exponential boundedness* of $g \circ \tau$.

But this leads immediately to assertion b), too: According to $\rho < 1$ follows

$$\begin{aligned}
E_\vartheta(g \circ \tau)^n &= \sum_{j=1}^{\infty} j^n \, P_\vartheta(g \circ \tau = j) \\
&\leq \sum_{j=1}^{\infty} j^n \, P_\vartheta(g \circ \tau \geq j) \overset{a)}{\leq} R \sum_{j=1}^{\infty} j^n \, \rho^j < \infty. \qquad \square
\end{aligned}$$

This theorem yields, moreover, the existence of all moments of the expected sampling costs:

(3.10) Corollary
    Under the assumption of (3.9) one obtains

$$E_\vartheta(C(\tau))^n < \infty \qquad \forall n \in \text{IN}.$$

Proof: From $c(\tau_i) \leq \max_{1 \leq j \leq m} c(j)$ follows

$$\begin{aligned}
E_\vartheta(C(\tau))^n &= E_\vartheta(\sum_{i=1}^{h(\tau)} c(\tau_i))^n \\
&\leq E_\vartheta(g(\tau) \max_{1 \leq j \leq m} c(j))^n, \text{ since } h(\tau) \leq g(\tau) \\
&= (\max_{1 \leq j \leq m} c(j))^n \, E_\vartheta(g(\tau))^n < \infty. \qquad \square
\end{aligned}$$

For the special case of bounded $\hat{t}$ theorem (3.9)a) yields also (3.6). Considering for the general case in the proof of (3.9) the intersection with the set $\{\tau_i \leq m \; \forall i \in \text{IN}\}$ one additionally obtains that for arbitrary $m \in \text{IN}$

$$P_\vartheta(0 < \tau_i \leq m \quad \forall i \in \text{IN}) = 0.$$

The methods of computing characteristics of SPRT's (e.g. [Eg], [W/R]) often make use of the fact that the logarithmic likelihood ratios form a homogeneous Markov process.

This property holds true also for SPPRT's:

(3.11) Theorem
For[26] $\hat{t} : \mathrm{IR} \to \mathrm{IN}_0$ and

$$\tau_1 := \hat{t}(1), \ \tau_{n+1} := \hat{t}(q_{(\tau_1,\ldots,\tau_n)}(\vartheta_2,\vartheta_1)), \ n \in \mathrm{IN}$$

the process $(Y_n)_{n\in\mathrm{IN}}$ defined by

$$Y_n := \log(q_{(\tau_1,\ldots,\tau_n)}(\vartheta_2,\vartheta_1)(X_1,\ldots,X_{\tau_1+\ldots+\tau_n}))$$

forms a homogeneous Markov process with transition kernel

$$K(B,y) = P_\vartheta(y + \sum_{i=1}^{\hat{t}(y)} \log(f(\vartheta_2,\vartheta_1)(X_i) \in B).$$

Proof: Consider $y,y_1,\ldots,y_n \in \mathrm{IR}, \ t_n := \tau_1 + \ldots + \tau_n$; using the notation $W_i := \log f(\vartheta_2,\vartheta_1)(X_i)$ one obtains

$$P_\vartheta(Y_{n+1} \le y \mid Y_1 = y_1,\ldots,Y_n = y_n) =$$

$$= \sum_{k=0}^{\infty} P_\vartheta(Y_n + \sum_{i=t_n+1}^{t_n+\hat{t}(Y_n)} W_i \le y \mid Y_1 = y_1,\ldots,Y_n = y_n, \ t_n = k) \cdot$$

$$\cdot P_\vartheta(t_n = k \mid Y_1 = y_1,\ldots,Y_n = y_n)$$

$$= \sum_{k=0}^{\infty} P_\vartheta(y_n + \sum_{i=k+1}^{k+\hat{t}(y_n)} W_i \le y \mid Y_1 = y_1,\ldots,Y_n = y_n, \ t_n = k)$$

$$\cdot P_\vartheta(t_n = k \mid Y_1 = y_1,\ldots,Y_n = y_n).$$

For fixed k

$$\{y_n + \sum_{i=k+1}^{k+\hat{t}(y_n)} W_i \le y\} \in \sigma(X_{k+1},\ldots)$$

and

$$\{Y_1 = y_1,\ldots,Y_n = y, \ t_n = k\} \in \sigma(X_1,\ldots,X_k);$$

therefore these events are independent and it follows

$$P_\vartheta(Y_{n+1} \le y \mid Y_1 = y_1,\ldots,Y_n = y_n) =$$

$$= \sum_{k=0}^{\infty} P_\vartheta(y_n + \sum_{i=k+1}^{k+\hat{t}(y_n)} W_i \le y) \cdot P_\vartheta(t_n = k \mid Y_1 = y_1 \ldots,Y_n = y_n)$$

$$= P_\vartheta(y_n + \sum_{i=1}^{\hat{t}(y_n)} W_i \le y),$$

since the X_i are independent and identically distributed. Analogously one obtains

---

[26]The condition $\hat{t}(x) = 0 \quad \forall x \notin (k_1, k_2)$ is not needed for this proposition.

68

$$P_\vartheta(Y_{n+1} \leq y \mid Y_n = y_n) =$$

$$= \sum_{k=0}^{\infty} P_\vartheta\Big(y_n + \sum_{i=k+1}^{k+\hat{t}(y_n)} W_i \leq y \mid Y_n = y_n, \ t_n = k\Big) \cdot P_\vartheta(t_n = k \mid Y_n = y_n)$$

$$= \sum_{k=0}^{\infty} P_\vartheta\Big(y_n + \sum_{i=k+1}^{k+\hat{t}(y_n)} W_i \leq y\Big) P_\vartheta(t_n = k \mid Y_n = y_n)$$

$$= P_\vartheta\Big(y_n + \sum_{i=1}^{\hat{t}(y_n)} W_i \leq y\Big),$$

i.e. altogether the assertion of theorem (3.11). $\qquad\qquad\square$

In particular it follows

(3.12) Corollary

Let (τ, φ) be an SPPRT (as in (3.5)); then the states $y \notin (\log k_1, \log k_2)$ of the Markov process

$$(\log(q_{(\tau_1,\ldots,\tau_n)}(\vartheta_2, \vartheta_1)(X_1, \ldots, X_{\tau_1 + \ldots + \tau_n})))_{n \in \mathbb{N}}$$

are absorbing whereas the states $y \in (\log k_1, \log k_2)$ are transient.

For a proof one has only to mention that the states $y \notin (\log k_1, \log k_2)$ are absorbing because of $\hat{t}(x) = 0 \quad \forall x \notin (k_1, k_2)$ and that the transience of $y \in (\log k_1, \log k_2)$ follows from (3.6). $\qquad\qquad\square$

But there are also essential differences in the Markov structures of purely sequential and sequentially planned probability ratio tests: For SPRT's $\hat{t}(x) = 1$ holds $\forall x \in (k_1, k_2)$; therefore one obtains the transition kernel

$$K_1(B, y) = P_\vartheta(y + \log(f(\vartheta_2, \vartheta_1)(X_1)) \in B).$$

This kernel is translation invariant

$$K_1(B, y) = K_1(B - y, \ 0),$$

(the increment of the process is independent of the present state). But it is just a characteristic feature of sequentially planned decision procedures that the size of the next sub-sample and therefore the distribution of the increment is determined according to the present state. This has consequences in particular for methods/algorithms to compute characteristics of SPPRT's (see § 4).

Finally, we are now able to give a *proof of remark (3.6):* For each $\vartheta \in \Theta$ the $\log f(\vartheta_2, \vartheta_1)(X_i)$ are independent identically distributed (non-degenerate) random variables. Therefore for each $\varepsilon > 0$ there exists an $m \in \mathbb{N}$ such that

$$P_\vartheta(\mid \sum_{i=1}^{n} \log f(\vartheta_2, \vartheta_1)(X_i) \mid > \log k_2 - \log k_1) > 1 - \varepsilon \quad \forall n \geq m$$

– else there would exist an $\varepsilon > 0$ s.t.

$$\limsup_{n \to \infty} P_\vartheta(\mid \sum_{i=1}^{n} \log f(\vartheta_2, \vartheta_1)(X_i) \mid \leq (\log k_2 - \log k_1)) \geq \varepsilon,$$

which according to theorem 2.9 of Doob ([Do], p. 121) has as a consequence that $\sum_{i=1}^{n} \log f(\vartheta_2, \vartheta_1)(X_i)$ is a.s. centered convergent where the centering constants may, because of the iid-property, be chosen independent of i; but since the characteristic function of $\log f(\vartheta_2, \vartheta_1)(X_i)$ is < 1 in a neighborhood of 0 this would lead to a contradiction.

We have

$$\{\tau \notin A\} \subset \{0 < \tau_i \leq m \quad \forall i \in \mathrm{IN}\} + \sum_{k=1}^{\infty} \{0 < \tau_i \leq m, 1 \leq i < k;\ \tau_k > m,\ \tau_{k+1} > 0\}.$$

Then on the one hand

$$P_\vartheta(0 < \tau_i \leq m \quad \forall i \in \mathrm{IN}) = 0,$$

and on the other hand it follows, using the Y_n of (3.11),

$$P_\vartheta(\tau_{k+1} > 0 \mid 0 < \tau_i \leq m,\ 1 \leq i < k;\ \tau_k > m)$$
$$\overset{(3.11)}{=} P_\vartheta(Y_k \in (\log k_1; \log k_2) \mid \log Y_{k-1} \in \{\hat{t} > m\}),$$
$$\text{since } \tau_k \text{ depends on } (Y_1, \ldots, Y_{k-1}) \text{ only by } Y_{k-1}$$
$$\overset{(3.11)}{=} P_\vartheta(Y_1 \in (\log k_1; \log k_2) \mid \log Y_0 \in \{\hat{t} > m\}),$$

where as in the proof of (3.9)

$$P_\vartheta(Y_1 \notin (\log k_1; \log k_2) \mid \log Y_0 \in \{\hat{t} > m\})$$
$$\geq \inf_{n > m} P_\vartheta(\mid \sum_{i=1}^{n} \log f(\vartheta_2, \vartheta_1)(X_i) \mid > \log k_2 - \log k_1)$$
$$\geq 1 - \varepsilon.$$

Hence it follows for each $k \in \mathrm{IN}$ that

$$P_\vartheta(0 < \tau_i \leq m,\ 1 \leq i < k,\ \tau_k > m,\ \tau_{k+1} > 0) < \varepsilon \cdot P_\vartheta(0 < \tau_i \leq m,\ 1 \leq i < k,\ \tau_k > m)$$

and therefore

$$P_\vartheta(\sum_{k=1}^{\infty} \{0 < \tau_i \leq m, 1 \leq i < k,\ \tau_k > m,\ \tau_{k+1} > 0\})$$
$$\leq \varepsilon \sum_{k=1}^{\infty} P_\vartheta(0 < \tau_i \leq m,\ 1 \leq i < k,\ \tau_k > m) \leq \varepsilon.$$

Since $\varepsilon > 0$ is arbitrary the assertion of (3.6) follows. $\qquad\square$

§ 4 Algorithms for computing the OC- and ASC-function of SPPRT's in the iid case

For the computation of the OC- and ASN-function of SPRT's a series of algorithms was constructed (see [D/U 1], [D/U 2], [E/M], [Eg], [W/R]). Our goal is to develop corresponding methods for computing characteristics (e.g. OC- and ASC-function) of SPPRT's. Our assumptions are similar to those which allow the construction of efficient algorithms for SPRT's (see in particular [Eg]):

(3.13) Assumptions

   (i) X_1 *is* $\mathbf{Z}$*-valued*

   (ii) $\log q_{(n)}(\vartheta_2, \vartheta_1)(X_1, \ldots, X_n) = \gamma_1(\vartheta_1, \vartheta_2) \sum_{i=1}^{n} X_i - \gamma_0(\vartheta_1, \vartheta_2)n$
      where $\gamma_1(\vartheta_1, \vartheta_2) \neq 0$,

   (iii) $\gamma_0(\vartheta_1, \vartheta_2)/\gamma_1(\vartheta_1, \vartheta_2) = g_0/g_1$ *where* $g_0, g_1 \in \mathbf{Z}$ *and* $\gamma_1(\vartheta_1, \vartheta_2)/g_1 \geq 0$.

Assumption (i) means that the possible values of X_i are integer multiples of a common "unit" – using computers an assumption of this kind seems to be inevitable. Assumption (ii) is fulfilled e.g. if the densities f_ϑ (with respect to the counting measure) belong to an exponential family of the form

$$f_\vartheta(x) = h(x) \, \exp(\zeta(\vartheta)x - \eta(\vartheta));$$

in this case

$$\log q_{(n)}(\vartheta_2, \vartheta_1)(X_1, \ldots, X_n) =$$
$$= \sum_{i=1}^{n}((\zeta(\vartheta_2) - \zeta(\vartheta_1))X_i - (\eta(\vartheta_2) - \eta(\vartheta_1)))$$
$$= (\zeta(\vartheta_2) - \zeta(\vartheta_1)) \sum_{i=1}^{n} X_i - (\eta(\vartheta_2) - \eta(\vartheta_1))n.$$

In particular, some "standard" classes of distributions fulfill these assumptions:

(3.14) Examples

  a) Binomial distributions: $Q_\vartheta = \mathcal{B}(m, \vartheta)$, $\vartheta \in (0, 1)$, $m \in \mathrm{IN}$ fixed. From

$$f_\vartheta(x) = \binom{m}{x} \vartheta^x (1 - \vartheta)^{m-x} = \binom{m}{x} \exp(x \log(\vartheta/(1 - \vartheta)) - m \log(1/(1 - \vartheta)))$$

    follows that the assumptions (i) and (ii) are fulfilled with
      $\gamma_1(\vartheta_1, \vartheta_2) = \log(\vartheta_2(1 - \vartheta_1)/\vartheta_1(1 - \vartheta_2))$,
      $\gamma_0(\vartheta_1, \vartheta_2) = m \log((1 - \vartheta_1)/(1 - \vartheta_2))$.

  b) Poisson distributions: $Q_\vartheta = \mathcal{P}(\vartheta)$, $\vartheta > 0$. Because of

$$f_\vartheta(x) = \frac{\vartheta^x}{x!} \exp(-\vartheta) = \frac{1}{x!} \exp(x \log \vartheta - \vartheta)$$

    the assumptions (i) and (ii) are fulfilled with
      $\gamma_1(\vartheta_1, \vartheta_2) = \log(\vartheta_2/\vartheta_1)$, $\gamma_0(\vartheta_1, \vartheta_2) = \vartheta_2 - \vartheta_1$.

  c) Negative binomial distributions: $Q_\vartheta = Nb(m, \vartheta)$, $\vartheta \in (0, 1)$,
    $m \in \mathrm{IN}$ fixed. From

$$f_\vartheta(x) = \binom{m + x - 1}{x} \vartheta^m (1 - \vartheta)^x = \binom{m + x - 1}{x} \exp(x \log(1 - \vartheta) - m \log 1/\vartheta)$$

follows that the assumptions (i) and (ii) are fulfilled with
$$\gamma_1(\vartheta_1, \vartheta_2) = \log((1 - \vartheta_2)/(1 - \vartheta_1)), \quad \gamma_0(\vartheta_1, \vartheta_2) = n \log(\vartheta_1/\vartheta_2).$$
$\square$

At first sight, assumption (iii) seems to be much more stringent; but for computer algorithms a restriction to rational numbers is inevitable. In Appendix B we describe how to compute, for a given real number, an appropriate rational approximation by continued fraction expansion and, moreover, we give a corresponding module[27].

Under the assumptions (3.13) one obtains

$$\log q_{(n)}(\vartheta_2, \vartheta_1)(X_1, \ldots, X_n) = \frac{\gamma_1(\vartheta_1, \vartheta_2)}{g_1}(g_1 \sum_{i=1}^{n} X_i - g_0\, n),$$

where the random variables

$$Z_n := g_1 \sum_{i=1}^{n} X_i - g_0\, n$$

are *integers* (i.e. the values of $\log q_{(n)}(\vartheta_2, \vartheta_1)$ are concentrated on a lattice). An SPPRT (τ, φ) may completely be described by the Z_n:

$$\varphi_a = 1_{\{Z_{g(a)} > 0\}}\, , \quad \tau_{n+1} = \tilde{t}(Z_n)$$

where $\tilde{t}(z) := \hat{t}(\exp(\frac{\gamma_1(\vartheta_1, \vartheta_2)}{g_1} z))$; the stopping bounds for $\tilde{t}$ are

$$c_1 := \frac{g_1}{\gamma_1(\vartheta_1, \vartheta_2)} \log k_1\, , \quad c_2 := \frac{g_1}{\gamma_1(\vartheta_1, \vartheta_2)} \log k_2.$$

Since the number of possible values of Z_n in $(c_1; c_2)$ is *finite* – namely just the number of integers within that interval – one may assume that $\tilde{t}$ is bounded. Then the assumption of (3.9) is fulfilled and it follows:

(3.15) Corollary
  Let the assumptions (3.13) be fulfilled; then one obtains for each SPPRT (τ, φ)

a) $E_\vartheta(g \circ \tau)^n < \infty$ $\forall n \in \mathrm{IN}$,

b) $E_\vartheta(C \circ \tau)^n < \infty$ $\forall n \in \mathrm{IN}$.

Moreover, theorem (3.11) yields that

$$\left(Z_{\tau_1 + \ldots + \tau_n}\right)_{n \in \mathrm{IN}}$$

is a homogeneous Markov chain (with state space $\mathbf{Z}$) where according to (3.12) the states $j \leq c_1$ and $j \geq c_2$ are absorbing and those with $c_1 < j < c_2$ are transient. This Markov chain may be described by the transition probabilities

$$\tilde{q}_{ij} := P_\vartheta(Z_{\tau_1 + \ldots + \tau_n} = j \mid Z_{\tau_1 + \ldots + \tau_{n-1}} = i)$$

---

[27]But even for SPRT's the consequences of such approximations seem to be unknown.

72

where

$$\tilde{q}_{ij} = \begin{cases} 1 & j = i \\ & \text{for} \qquad , \text{if } i \notin (c_1, c_2) \\ 0 & j \neq i \end{cases}$$

and

$$\tilde{q}_{ij} = P_\vartheta(Z_{\tilde{t}(i)} = j - i)$$
$$= \begin{cases} P_\vartheta(\sum_{k=1}^{\tilde{t}(i)} X_k = \frac{j-i+g_0\tilde{t}(i)}{g_1}) & \text{for } j - i + g_0\tilde{t}(i) \equiv 0 \mod |g_1| \\ 0 & \text{elsewhere.} \end{cases}$$

if $i \in (c_1, c_2)$. For many characteristics of SPPRT's (e.g. OC-, PF-, ASC-function and, moreover, moments of $g \circ \tau$) the only important property of the absorbing states is whether they are $\leq c_1$ or $\geq c_2$. Therefore the structure of the Markov chain may further be simplified by collecting all states $\leq c_1$ to one new state and all states $\geq c_2$ to a second new state. For the algorithms developed in the sequel we therefore consider a homogeneous Markov chain with state space

$$E := \{1, \ldots, m, m+1, m+2\}$$

where $m := \lceil c_2 \rceil - \lfloor c_1 \rfloor - 1$ is the number of integers between c_1 and c_2; the states $1, \ldots, m$ in E correspond to the states $\lfloor c_1 \rfloor + 1, \ldots, \lceil c_2 \rceil - 1$ of the original chain, the state $m+1$ is the union of all original states[28] $\geq c_2$, and the state $m+2$ is the union of all original states $\leq c_1$. This Markov chain has the transition probabilities

$$q_{ij} = \begin{cases} \delta_{ij} & \text{for } i \in F := \{m+1, \, m+2\} \\ P_\vartheta(Z_{\tilde{t}(\lfloor c_1 \rfloor + i)} = j - i) & \text{for } i, j \notin F \\ P_\vartheta(Z_{\tilde{t}(\lfloor c_1 \rfloor + i)} \geq m + 1 - i) & \text{for } i \notin F, \, j = m+1 \\ P_\vartheta(Z_{\tilde{t}(\lfloor c_1 \rfloor + i)} \leq -i) & \text{for } i \notin F, \, j = m+2, \end{cases}$$

i.e. in particular $q_{ij} = \tilde{q}_{\lfloor c_1 \rfloor + i, \lfloor c_1 \rfloor + j}$ for $i, j \notin F$.

Using the notation $t(i) := \tilde{t}(\lfloor c_1 \rfloor + i)$ and denoting by F_n the distribution function of $\sum_{k=1}^{n} X_k$ (i.e. of Q_ϑ^{*n}) one therefore obtains:

(3.16) Remark
(i) If $i, j \notin F$, then

$$q_{ij} = \begin{cases} P_\vartheta(\sum_{k=1}^{t(i)} X_k = \frac{j-i+g_0t(i)}{g_1}), & \text{if } j - i + g_0t(i) \equiv 0 \,(\mathrm{mod}\,|g_1|) \\ 0 & \text{elsewhere} \end{cases}$$

---

[28]This somewhat artificial notation allows a simple distinction between transient ($i \leq m$) and absorbing ($i > m$) states.

(ii) If $i \notin F$ then

$$q_{i,m+1} = \begin{cases} P_\vartheta(\sum_{k=1}^{t(i)} X_k \geq \frac{m+1-i+g_0 t(i)}{g_1}), & \text{if } g_1 > 0 \\ P_\vartheta(\sum_{k=1}^{t(i)} X_k \leq \frac{m+1-i+g_0 t(i)}{g_1}), & \text{if } g_1 < 0 \end{cases}$$

$$= \begin{cases} 1 - F_{t(i)}(\frac{m-i+g_0 t(i)}{g_1}), & \text{if } g_1 > 0 \\ F_{t(i)}(\frac{m+1-i+g_0 t(i)}{g_1}), & \text{if } g_1 < 0. \end{cases}$$

(iii) If $i \notin F$ then

$$q_{i,m+2} = \begin{cases} P_\vartheta(\sum_{k=1}^{t(i)} X_k \leq \frac{-i+g_0 t(i)}{g_1}), & \text{if } g_1 > 0 \\ P_\vartheta(\sum_{k=1}^{t(i)} X_k \geq \frac{-i+g_0 t(i)}{g_1}), & \text{if } g_1 < 0 \end{cases}$$

$$= \begin{cases} F_{t(i)}(\frac{-i+g_0 t(i)}{g_1}), & \text{if } g_1 > 0 \\ 1 - F_{t(i)}(\frac{-i+1+g_0 t(i)}{g_1}), & \text{if } g_1 < 0. \end{cases}$$

If the convolutions Q_ϑ^{*n} are known in closed form – as e.g. for binomial-, Poisson- or negative binomial-distributions – the q_{ij} can be given in a simple way; otherwise one has to compute these values numerically.

The dimension of the matrix

$$\overline{Q} := (q_{ij})_{1 \leq i,j \leq m+2},$$

which belongs to an SPPRT (τ, φ), is

$$m + 2 \approx \lceil c_2 \rceil - \lfloor c_1 \rfloor + 1 \approx \frac{g_1}{\gamma_1(\vartheta_1, \vartheta_2)}(\log k_2 - \log k_1).$$

Since this matrix is of major importance for the computation of characteristics of (τ, φ) one has, for reasons of efficiency, to take care that the denominator $|g_1|$ of the rational approximation g_0/g_1 of $\gamma_0(\vartheta_1, \vartheta_2)/\gamma_1(\vartheta_1, \vartheta_2)$ is not "too large":

(3.17) Example

Let $Q_\vartheta = \mathcal{B}(1, \vartheta)$, $\vartheta \in (0, 1)$, $\vartheta_1 = 0.6$, $\vartheta_2 = 0.7$ and (τ, φ) be an SPPRT which uses the stopping bounds of (Wald's) approximation (3.7)c) for $\alpha = \beta = 0.1$ i.e. $k_1 = 1/9$, $k_2 = 9$. Then

$$\log k_2 - \log k_1 = 4.39445$$

and (according to (3.14)a))

$$\gamma_1(\vartheta_1, \vartheta_2) = 0.441833\ldots \quad , \quad \gamma_0(\vartheta_1, \vartheta_2) = 0.287682\ldots,$$
$$\gamma_0(\vartheta_1, \vartheta_2)/\gamma_1(\vartheta_1, \vartheta_2) = 0.651111\ldots$$

If $\varepsilon = 10^{-3}$ is required for the accuracy of rational approximation then certainly $g_0' = 651, g_1' = 1000$ yield an obvious admissible approximation leading to a transition matrix of size 9947. But also the pair $g_0 = 28$, $g_1 = 43$ fulfills the accuracy requirement $(|\gamma_0(\vartheta_1, \vartheta_2)/\gamma_1(\vartheta_1, \vartheta_2) - g_0/g_1| = 5.28 \cdot 10^{-5})$; the corresponding matrix has size 429 (in Appendix B it is shown that this approximation has the smallest denominator among all fulfilling the accuracy requirement and thus leads to the "smallest"

74

transition matrix). Under computational aspects therefore the choice of a reasonable rational approximation of $\gamma_0(\vartheta_1, \vartheta_2)/\gamma_1(\vartheta_1, \vartheta_2)$ is essential (comp. Appendix B).

For the development of efficient algorithms for SPPRT's furthermore a structural property ("sparseness") of the transition matrix $\overline{Q}$ turns out to be important:

(3.18) Remark

According to (3.16) the transition probabilities q_{ij} of the transient states $i, j \le m$ vanish for

$$j - i + g_0 t(i) \not\equiv 0 \pmod{\mid g_1 \mid}.$$

Therefore only a proportion of about $1/\mid g_1 \mid$ of the elements of $\overline{Q}$ is non-trivial. This can be used for an efficient storage of $\overline{Q}$; the following scheme seems to be reasonable: For each row $i \le m$ an offset

$$\begin{aligned} O_i \; &:= \; \min\{j \ge 1 : j - i + g_0\, t(i) \equiv 0 \;\; \mathrm{mod} \;\; \mid g_1 \mid\} \\ &= \; [(-(1 - i + g_0\, t(i))) \;\; \mathrm{mod} \;\; \mid g_1 \mid] + 1 \end{aligned}$$

is defined and only the values

$$q_{i, O_i + (k-1)\mid g_1 \mid} \qquad (k = 1, \ldots, l_i)$$

where $l_i := \max\{k' \in \mathbb{N}_0 : 1 \le O_i + (k' - 1) \mid g_1 \mid \le m\}$ will be stored ($\max \emptyset \le 0$ arbitrary). One obtains:

$$l_i = \left[\frac{m - O_i}{\mid g_1 \mid} \right] + 1.$$

For analyzing execution times and store requirements one may assume $l_i \approx m/\mid g_1 \mid$. Using this kind of storage it is sufficient to store only $\frac{1}{\mid g_1 \mid} m^2 + 2m$ values ($+\, m$ additional informations) instead of m^2 items. In example (3.17) with $g_1 = 43$ and $m = 427$ one has e.g. to store only 5522 values instead of 182 329.

It should be mentioned that for the special case of SPRT's a further simplification of the matrix $\overline{Q}$ is possible: In this case $t(i) = 1$, $1 \le i \le m$. Hence for $i, j \le m$ the transition probabilities q_{ij} vanish for

$$j - i \not\equiv -g_0 \pmod{\mid g_1 \mid},$$

i.e. the non-trivial entries are concentrated on side-diagonals; moreover the values of such a diagonal coincide. This shows that for the purely sequential case the matrix $\overline{Q}$ is much simpler than for the sequentially planned case. Consequently, only a part of the algorithms developed for SPRT's allow a generalization to the case of sequentially planned tests.

Besides saving storage space, the implementation of the matrix $\overline{Q}$ as mentioned above allows speed improvement since many additions of and multiplications by 0 are saved:

(3.19) Examples

a) Computing $y = \sum_{j=1}^{m} q_{ij} x_j$ is realized by

$y := 0$;
FOR $k := 1$ TO l_i DO

$$y := y + q[i, O_i + (k-1) \mid g_1 \mid] \star x[O_i + (k-1) \mid g_1 \mid]$$

END;

This algorithm carries out about $m/\mid g_1 \mid$ operations compared to m operations needed without using the special structure of $\overline{Q}$. Computing $\overline{Q}x$ then needs about $m^2/\mid g_1 \mid$ operations instead of m^2 operations.

b) Computing $y = (y_1, \ldots, y_m)$ where $y_j := \sum_{i=1}^{m} q_{ij} x_i$ is realized by

FOR $i := 1$ TO m DO

$y[i] := 0$;

END;
FOR $i := 1$ TO m DO
  FOR $k := 1$ TO l_i DO

$$y[O_i + (k-1) \mid g_1 \mid] :=$$
$$y[O_i + (k-1) \mid g_1 \mid] + q[i, O_i + (k-1) \mid g_1 \mid] \star x[i]$$

  END;
END;

This algorithm carries out about $m^2/\mid g_1 \mid$ operations instead of m^2 operations needed without using the special structure of $\overline{Q}$. $\quad\square$

In the sequel we develop algorithms for the computation of
- the operating characteristic

$$OC(\vartheta) = P_\vartheta(Z_{g(\tau)} \leq c_1) = E_\vartheta(\sum_{i=1}^{h(\tau)} 1_{(-\infty;c_1]}(Z_{\tau_1+\ldots+\tau_i}))$$

- the power function

$$PF(\vartheta) = P_\vartheta(Z_{g(\tau)} \geq c_2) = E_\vartheta(\sum_{i=1}^{h(\tau)} 1_{[c_2;\infty)}(Z_{\tau_1+\ldots+\tau_i}))$$

- the ASC-function

$$\begin{aligned} ASC(\vartheta) &= E_\vartheta\, C(\vartheta, \tau) \\ &= E_\vartheta(\sum_{i=1}^{h(\tau)} c(\tau_i)) = E_\vartheta(\sum_{i=0}^{h(\tau)} c(\tilde{t}(Z_{\tau_1+\ldots+\tau_i}))). \end{aligned}$$

All these functions – and moreover e.g. also the expected total sample size

$$E_\vartheta(g \circ \tau) = E_\vartheta(\sum_{i=1}^{h(\tau)} \tau_i) \quad -$$

76

are of the form

$$E_\vartheta(\sum_{i=0}^{h(\tau)} W(Z_{\tau_1+\dots+\tau_i})),$$

where $(Z_{\tau_1+\dots+\tau_n})_{n\in\mathbb{N}}$ is a homogeneous Markov chain (with a special structure). Therefore we consider a homogeneous Markov chain $(Y_n)_{n\in\mathbb{N}_0}$ with state space

$$E = \{1,\dots,m+2\}$$

where the states $1,\dots,m$ are transient and the states $i \in F = \{m+1, m+2\}$ are absorbing; the matrix of the transition probabilities

$$q_{ij} = P_\vartheta(Y_n = j \mid Y_{n-1} = i)$$

is denoted by $\overline{Q}$, the matrix of the transient states by[29] Q, i.e.

$$Q = (q_{ij})_{1 \le i,j \le m}.$$

For this Markov chain stopping rules of the form

$$N = \inf\{n \in \mathbb{N}_0 : Y_n \in F\}$$

or

$$N_M := \inf\{n \in \mathbb{N}_0 : Y_n \in F \text{ or } n \ge M\}$$

are considered (the N_M are of interest in particular for approximations using finite horizons) and, moreover, a "weighting" function

$$W : E \to \mathbb{R}.$$

According to our previous remark the aim is to compute

$$E_\vartheta(\sum_{i=0}^{N} W(Y_i)).$$

(I) The LE-method

For this situation there exists a standard method for computing the probability of absorption, leading to a system of linear equations (the algorithm based on this method will, therefore, be called LE-method). Let

$$\mu_i^{(k)} := E_\vartheta((\sum_{h=0}^{N} W(Y_h))^k \mid Y_0 = i), \ k \in \mathbb{N}_0, \ i \le m,$$

(i.e. if i is the initial value of the Markov chain then $\mu_i^{(1)}$ is the expectation one is interested in)[30]

$$\begin{aligned}
\mu^{(k)} &:= (\mu_1^{(k)},\dots,\mu_m^{(k)})^t \\
q_{ij}^{(k,l)} &:= q_{ij}W(i)^{k-l} \ , \ l < k, \\
\tilde{Q}^{(k,l)} &:= (q_{ij}^{(k,l)})_{1 \le i,j \le m}, \ l < k, \\
d_i^{(k)} &:= \sum_{l=0}^{k} \binom{k}{l} W(i)^{k-l} \sum_{j>m} q_{ij}\, W(j)^l,
\end{aligned}$$

---

[29] The spectral radius $\rho(Q)$ of Q is < 1.

[30] There is no connection between these $q_{ij}^{(k,l)}$ and the k-step transition probabilities which are often denoted in a similar way.

and

$$d^{(k)} := (d_1^{(k)}, \ldots, d_m^{(k)})^t.$$

Using these notations we are able to formulate

(3.20) Lemma
$\mu^{(0)} = (1, \ldots, 1)^t$ and

$$(I - Q)\,\mu^{(k)} = d^{(k)} + \sum_{l=0}^{k-1} \binom{k}{l} \tilde{Q}^{(k,l)} \mu^{(l)} \quad for\ k \in \mathbb{N}.$$

Proof: Consider $k \in \mathbb{N}$ and $i \leq m$; according to the definition of $\mu_i^{(k)}$ one obtains from the binomial formula

$$
\begin{aligned}
\mu_i^{(k)} &= E_\vartheta\big(\sum_{l=0}^{k} \binom{k}{l} W(Y_0)^{k-l} (\sum_{h=1}^{N} W(Y_h))^l \mid Y_0 = i\big) \\
&= \sum_{l=0}^{k} \binom{k}{l} W(i)^{k-l} E_\vartheta\big((\sum_{h=1}^{N} W(Y_h))^l \mid Y_0 = i\big).
\end{aligned}
$$

Splitting up the remaining expectation according to the value of Y_1 yields

$$\sum_{j=1}^{m+2} \underbrace{E_\vartheta\big((\sum_{h=1}^{N} W(Y_h))^l \mid Y_0 = i,\ Y_1 = j\big)}\ P_\vartheta(Y_1 = j \mid Y_0 = i)$$

$$= \begin{cases} W(j)^l, & \text{if } j > m \ (\text{since } N = 1) \\[2mm] \mu_j^{(l)} & \text{elsewhere (since } (Y_n)_{n \in \mathbb{N}_0} \text{ is a Markov-chain)} \end{cases} ;$$

and therefore

$$
\begin{aligned}
\mu_i^{(k)} &= \sum_{l=0}^{k} \binom{k}{l} W(i)^{k-l} \sum_{j>m} q_{ij} W(j)^l + \\
&\quad + \sum_{l=0}^{k-1} \binom{k}{l} \sum_{j=1}^{m} q_{ij}\, W(i)^{k-l}\, \mu_j^{(l)} + \sum_{j=1}^{m} q_{ij}\, \mu_j^{(k)},
\end{aligned}
$$

i.e.

$$\mu^{(k)} = d^{(k)} + \sum_{l=0}^{k-1} \binom{k}{l} \tilde{Q}^{(k,l)} \mu^{(l)} + Q\mu^{(k)}. \qquad \square$$

Because of $\rho(Q) < 1$ the matrix $I - Q$ may be inverted; hence the linear equations from (3.20) have a unique solution. Using the inverse of $I - Q$ one can, therefore, successively determine the moments $\mu^{(k)}$.

The next corollaries contain the concrete linear equations for the OC-, the PF- and the ASC-function of SPPRT's; the really interesting values are (due to the index-translation) obtained as

$$\mu_{-\lfloor c_1 \rfloor}^{(k)}.$$

(3.21) Corollary

Let $\mu^{OC} \in \mathbb{R}^m$ be the vector of the OC-values of the states $1 \le i \le m$ and $d_i^{OC} := q_{i,m+2}$. Then

$$(I - Q)\mu^{OC} = d^{OC}.$$

Proof: For $W(i) = 1_{\{m+2\}}(i)$ and $k = 1$ follows for $i \le m$

$$d_i^{(1)} = \sum_{l=0}^{1} \binom{k}{l} W(i)^{1-l} \sum_{j>m} q_{ij} W(j)^l = q_{i,m+2};$$

moreover

$$q_{ij}^{(1,0)} = q_{ij} \, W(i) = 0,$$

i.e. $\tilde{Q}^{(1,0)} = (0)$; (3.20) therefore yields the assertion. $\quad\square$

In an analogous way one obtains

(3.22) Corollary

Let $\mu^{PF} \in \mathbb{R}^m$ be the vector of the PF-values of the states $1 \le i \le m$ and $d_i^{PF} := q_{i,m+1}$. Then

$$(I - Q) \, \mu^{PF} = d^{PF}.$$

For the ASC-function follows

(3.23) Corollary

Let $\mu^{ASC} \in \mathbb{R}^m$ be the vector of the ASC-values of the states $1 \le i \le m$ and $d_i^{ASC} := c(t(i))$. Then

$$(I - Q) \, \mu^{ASC} = d^{ASC}.$$

Proof: For

$$W(i) = \begin{cases} c(t(i)) & \text{if } i \le m \\ 0 & \text{elsewhere} \end{cases}$$

and $k = 1$ follows for $i \le m$

$$d_i^{(1)} = W(i) \sum_{j>m} q_{ij} = c(t(i)) \sum_{j>m} q_{ij}$$

and

$$\sum_{j=1}^{m} q_{ij}^{(1,0)} \cdot 1 = \sum_{j=1}^{m} q_{ij} \, W(i) = \sum_{j=1}^{m} q_{ij} \, c(t(i)).$$

Since $\sum\limits_{i=1}^{m+2} q_{ij} = 1$ lemma (3.20) yields the assertion. $\quad\square$

As a further result we mention

(3.24) Corollary

Let $\mu^{VC} \in \mathbb{R}^m$ be the vector of second moments of $ASC(\vartheta)$ for the states $1 \le i \le m$ and

$$d_i^{VC} := c(t(i))^2 + 2 \, c(i) \sum_{j=1}^{m} q_{ij} \, \mu_j^{ASC}, \quad 1 \le i \le m;$$

then

$$(I - Q)\,\mu^{VC} = d^{VC}.$$

Proof: With $W(i)$ as in (3.23) and $k = 2$ follows for $i \leq m$

$$d_i^{(2)} = \sum_{l=0}^{2} \binom{2}{l} W(i)^{2-l} \sum_{j>m} q_{ij}\, W(j)^l$$

$$= 1 \cdot W(i)^2 \sum_{j>m} q_{ij}$$

$$\sum_{j=1}^{m} \binom{2}{0} q_{ij}^{(2,0)} \cdot 1 = \sum_{j=1}^{m} q_{ij}\, W(i)^2$$

$$\sum_{j=1}^{m} \binom{2}{1} q_{ij}^{(2,1)} \mu_j^{(1)} = \sum_{j=1}^{m} 2 q_{ij}\, W(i)\, \mu_j^{ASC},$$

hence together with (3.20) the assertion. $\qquad\square$

For SPPRT's where the corresponding Markov chain has only very few states these assertions may be used to compute the OC-, the PF- and the ASC-function in closed form (for an example of Lübbert [Lü] this is carried out by Meyerthole [Me]).

In the general case the inversion of the matrix $I - Q$ which is, according to (3.20), the main step for computing the above mentioned characteristics of SPPRT's can be performed by standard methods of numerical mathematics (e.g. Gaussian elimination). But even for a clever choice of $g_0(\vartheta_1, \vartheta_2)$ and $g_1(\vartheta_1, \vartheta_2)$ the matrix will, in general, be rather large. Then Gaussian elimination causes several problems: Firstly, for the Gaussian elimination one needs an LR-decomposition of $I - Q$ and this will, in general, no longer be sparse (i.e. the special structure of Q is not used and there arise storage problems). Secondly, the computational effort is very high for "large" matrices (about $m^3/3$ operations). Thirdly, using Gaussian elimination for large matrices (with possibly small elements) may lead to considerable rounding errors. Therefore this direct LE-method will not be implemented; but the fact that the interesting characteristics may be computed by solving a system of linear equations will be used for the next algorithm where the inversion of $I - Q$ is performed by an iteration procedure.

Well-known iterative methods for solving systems of linear equations are the Gauss-Seidel-procedure and successive overrelaxation (SOR). For the solution of $(I - Q)x = b$ the Gauss-Seidel-procedure is based on the iteration

$$x_i^{(\nu+1)} := \frac{1}{1 - q_{ii}}\Big(b_i + \sum_{j=1}^{i-1} q_{ij}\, x_j^{(\nu+1)} + \sum_{j=i+1}^{m} q_{ij}\, x_j^{(\nu)}\Big);\ 1 \leq i \leq m,$$

where $x^{(0)}$ is, in principle, arbitrary (but it should be not too far from the actual solution). To use this procedure one evidently needs the condition

$$q_{ii} \neq 1,\ 1 \leq i \leq m;$$

but this is fulfilled due to the transience of the states $i \leq m$. The SOR-procedure is a generalization of the previous method; it is based on the iteration

$$x_i^{(\nu+1)} := \omega\, \frac{1}{1 - q_{ii}}\Big(b_i + \sum_{j=1}^{i-1} q_{ij}\, x_j^{(\nu+1)} + \sum_{j=i+1}^{m} q_{ij}\, x_j^{(\nu)}\Big) + (1 - \omega)\, x_i^{(\nu)},$$

80

where $x^{(0)}$ is again arbitrary and $\omega \in (0, 2)$ is a parameter which (hopefully) ought to improve the speed of convergence ($\omega = 1$ just yields the Gauss-Seidel-procedure). Decomposing Q into

$$Q = L + D + R$$

where

$$L = \begin{pmatrix} 0 & \cdots & \cdots & & 0 \\ q_{2,1} & \ddots & & & \vdots \\ \vdots & \ddots & \ddots & & 0 \\ q_{m,1} & \cdots & q_{m,m-1} & & 0 \end{pmatrix}, \quad R = \begin{pmatrix} 0 & q_{1,2} & \cdots & q_{1,m} \\ \vdots & & & \\ \vdots & & & q_{m-1,m} \\ 0 & \cdots & \cdots & 0 \end{pmatrix}, \quad D = \begin{pmatrix} q_{11} & 0 & \cdots & 0 \\ 0 & q_{22} & & \vdots \\ \vdots & & \ddots & 0 \\ 0 & & \cdots & q_{mm} \end{pmatrix}$$

allows to rewrite the iteration of SOR in the form

$$x^{(\nu+1)} = \omega(I - D)^{-1}(b + Lx^{(\nu+1)} + Rx^{(\nu)}) + (1 - \omega)\, x^{(\nu)},$$

i.e.

$$(I - D - \omega L)\, x^{(\nu+1)} = \omega(b + R\, x^{(\nu)}) + (1 - \omega)\,(I - D)\, x^{(\nu)}$$

and therefore

$$x^{(\nu+1)} = \omega(I - D - \omega L)^{-1}\, b + (I - D - \omega L)^{-1}\, (\omega\, R + (1 - \omega)(I - D))x^{(\nu)}.$$

$(I - D - \omega L)$ is invertible since it is a lower triangular matrix with diagonal elements $\neq 0$. If the sequence $(x^{(\nu)})_{\nu \geq 0}$ converges then the limit solves the equation $(I - Q)\, x = b$. To prove convergence the following general assertion (resulting from Banach's fixed point theorem) turns out to be useful:

(3.25) Remark

Let B a complex matrix and let the sequence $(x^{(\nu)})_{\nu \geq 0}$ be defined by

$$x^{(\nu+1)} := B\, x^{(\nu)} + c.$$

$(x^{(\nu)})_{\nu \geq 0}$ converges for each $x^{(0)} \in \mathbf{C}^m$ and each $c \in \mathbf{C}^m$ iff the spectral radius $\rho(B)$ of B is smaller than 1; then the rate of convergence is geometrical and is determined by $\rho(B)$.

Investigating the convergence of SOR one is therefore interested in the spectral radius of the matrix

$$B_\omega := (I - D - \omega L)^{-1}\, (\omega\, R + (1 - \omega)(I - D)) :$$

(3.26) Theorem

Let $0 < \omega \leq 1$; then SOR converges and for $\omega \geq 1/2$

$$\rho(B_\omega) \leq 1 + \omega(\rho(Q) - 1).$$

In particular $\rho(B_1) \leq \rho(Q)$.

Proof: Define $r := \rho(Q) < 1$ and $\lambda := \rho(B_\omega)$. From $0 \leq D + \omega L \leq D + L \leq Q$ follows

$$\rho(D - \omega L) \leq r < 1$$

and therefore

$$(I - D - \omega L)^{-1} = \sum_{k \geq 0}(D + \omega L)^k \geq 0.$$

Hence $B_\omega \geq 0$ and there exists a $y \geq 0$ s.t.

$$B_\omega y = \lambda y.$$

This yields

$$(\omega R + (1 - \omega)(I - D))y = \lambda y - \lambda D y - \lambda \omega L y$$

and therefore

$$(\omega R + (\lambda + \omega - 1)D + \lambda \omega L)y = (\lambda + \omega - 1)y,$$

i.e. $\lambda + \omega - 1$ is an eigenvalue of $\omega R + (\lambda + \omega - 1)D + \lambda \omega L$; hence

$$\lambda + \omega - 1 \leq \rho(\omega R + (\lambda + \omega - 1)D + \lambda \omega L).$$

For $\lambda \geq 1$ it would follow that

$$0 \leq \omega \leq \lambda + \omega - 1$$

and

$$\lambda \omega \leq \lambda + \omega - 1, \quad \text{since } (1 - \omega)(\lambda - 1) \geq 0,$$

yielding

$$\omega R + (\lambda + \omega - 1)\,D + \lambda \omega L \leq (\lambda + \omega - 1)(R + D + L)$$

and therefore[31]

$$\lambda + \omega - 1 \leq \rho((\lambda + \omega - 1)(R + D + L)) = (\lambda + \omega - 1)r;$$

because of $\lambda + \omega - 1 \geq 0$ this leads to the contradiction $r \geq 1$. Hence $\lambda < 1$ and, according to (3.25), SOR converges.

Moreover, for $\omega \geq 1/2$ it follows that

$$
\begin{aligned}
| \,\omega R \;+\; & (\lambda + \omega - 1)\,D + \lambda \omega L \,| = \\
\leq \;\; & \omega R + \underbrace{|\,\lambda - 1 + \omega\,|}_{\leq |\lambda - 1/2| + |\omega - \frac{1}{2}| \leq \omega, \text{ since } \omega \geq 1/2} D + \lambda \omega L \\
\leq \;\; & \omega(R + D + L) = \omega Q.
\end{aligned}
$$

This yields[31]

$$\lambda + \omega - 1 \leq \omega \rho(Q),$$

i.e. the assertion $\lambda \leq 1 + \omega(\rho(Q) - 1)$. $\qquad\qquad\square$

(3.27) Theorem

  Let $1 \leq \omega < 2/(1 + \rho(Q))$. Then $\rho(B_\omega) < 1$ and the SOR-procedure converges.

Proof: For $r := \rho(Q)$ follows

$$\rho(D + \omega L) \leq \rho(\omega(D + L)) \leq \omega \rho(Q) = \omega r < 2r/(1 + r) < 1.$$

---

[31]If $A \geq 0$ and C are $(m \times m)$-matrices s.t. $|\,C\,| \leq A$ then $\rho(C) \leq \rho(A)$.

82

Therefore

$$(I - D - \omega L)^{-1} = \sum_{k \in \mathbb{N}_0} (D + \omega L)^k \geq 0$$

yielding

$$A_\omega := (I - D - \omega L)^{-1}(\omega R + (\omega - 1)(I - D)) \geq 0$$

and

$$\mid B_\omega \mid \leq A_\omega.$$

Because of $A_\omega \geq 0$ there exists, for $\lambda := \rho(A_\omega)$, a $y \neq 0$ s.t. $A_\omega \, y = \lambda y$; therefore one obtains from the definition of A_ω

$$(\omega R + (\lambda - \omega + 1)D + \omega \lambda L) \, y = (\lambda - \omega + 1)y,$$

i.e. $(\lambda - \omega + 1)$ is an eigenvalue of

$$\omega R + (\lambda - \omega + 1)D + \omega \lambda L;$$

hence

$$(\lambda - \omega + 1) \leq \rho(\omega R + (\lambda - \omega + 1)D + \omega \lambda L).$$

Assume that $\lambda \geq 1$. Then it would follow

$$\lambda - \omega + 1 \geq 0, \quad \text{since } \omega < 2/(1 + r) < 2,$$

and, on the other hand,

$$\lambda - \omega + 1 \leq \lambda \omega, \quad \text{since } 0 \leq (\omega - 1)(\lambda + 1).$$

From $\lambda \geq 1$ and footnote 31 one would therefore obtain

$$\lambda - \omega + 1 \leq \rho(\omega \lambda R + \omega \lambda D + \omega \lambda L) = \omega \lambda r,$$

hence

$$\lambda + 1 \leq \omega(1 + \lambda r)$$

and

$$\begin{aligned}
\omega \;\geq\;& (1 + \lambda)/(1 + \lambda r) \\
=\;& \frac{(1 + \lambda)(1 + r)}{(1 + \lambda r)(1 + r)} \\
=\;& \frac{2(1 + \lambda r) + (\lambda - 1)(1 - r)}{(1 + \lambda r)(1 + r)} \geq \frac{2}{1 + r}
\end{aligned}$$

which contradicts the assumption on ω. Therefore $\lambda < 1$ and it follows (again using footnote 31)

$$\rho(B_\omega) \leq \rho(A_\omega) = \lambda < 1. \qquad \square$$

(3.26) and (3.27) ensure the convergence of the SOR-algorithm for $0 < \omega < 2/(1 + \rho(Q))$.

To estimate $\rho(Q)$, standard methods are available e.g.

$$\rho(Q) \leq \|Q\|_\infty \leq \max_{1\leq i\leq m} \sum_{j=1}^{m} q_{ij},$$

where the last term is < 1 for all "reasonable" SPPRT's (starting from an arbitrary state an absorption is possible – else one would save fixed costs by enlarging the sub-samples) and therefore the SOR-algorithm converges for $\omega \in (0; 2/(1+\max_i \sum_j q_{ij}))$. But as usual for SOR-procedures the question arises how to choose ω in order to make $\rho(B_\omega)$ small (as small as possible). Since on the one hand the computation of a spectral radius is rather tedious and, on the other hand, there does not exist any general criterion of a favourable choice of ω, one will in many cases content oneself with the Gauss-Seidel procedure (i.e. $\omega = 1$). For this method one obtains (using Banach's fixed point theorem) simple error bounds:

(3.28) Remark

Let $\max_{1\leq i\leq m} \sum\limits_{j=1}^{m} q_{ij} < 1$*; then*

$$r := \max_{1\leq i\leq m} \sum_{\substack{j=1 \\ j\neq i}}^{m} q_{ij}/(1 - q_{ii}) < 1$$

and it follows
(i) $\quad \|x^{(\nu)} - x\|_\infty \leq \frac{r}{1-r}\|x^{(\nu)} - x^{(\nu-1)}\|_\infty$

(ii) $\|x^{(\nu)} - x\|_\infty \leq \frac{r^\nu}{1-r}\|x^{(1)} - x^{(0)}\|_\infty.$

The main advantage of Gauss-Seidel/SOR-algorithms consists in utilizing the special structure of Q: During the iterations one may use the algorithm (3.19)a) requiring about $m^2/\mid g_1 \mid$ operations per iteration (for an implementation comp. Appendix B). A further advantage of SOR is its robustness against rounding errors since (3.26)/ (3.27) guarantee convergence for all starting points.

A disadvantage is that one has to start a new iteration for each new "right hand side" of the linear equations (from (3.20)) – using Gaussian elimination one has only once to compute – an i.g. very complicated – LR-decomposition which in a simple way leads to solutions for different "right hand sides". But this disadvantage is more than compensated (due to efficient storage and smaller execution times) by utilizing the special structure of Q.

If one has to compute the characteristics for several parameter values $\vartheta_0 < \vartheta_1 < \ldots < \vartheta_k$ it seems to be promising to reduce the number of necessary iterations (for prescribed accuracy) in the following way: For ϑ_0 one chooses an arbitrary starting point, and for $i \geq 1$ one takes for ϑ_i as starting point just the solution obtained for ϑ_{i-1} (expecting that this value is, in many cases, a better approximation than a starting point chosen without any a priori information (as e.g. the null-vector)). In Appendix B this method will be called *ILE-method without new initialization* whereas always choosing the null-vector as starting point is called *ILE-method with new initialization* (for a comparison

84

see § 5).

(II) The BF-method

Our second method – used by Dieter/Unger [D/U 1], [D/U 2] in the special case of SPRT's for Bernoulli-distributions – consists in representing the characteristics (OC-, PF-, ASC-function) as series and summing these up by "brute force" until a prescribed accuracy is achieved (therefore this method will be called BF-method).

For a Markov chain with known initial distribution $p^{(0)}$ the distribution of $Y_n, n \in \mathbb{N}_0$, may be computed by recursion: Using the notations

$$
\begin{aligned}
p_j^{(i)} &:= P_\vartheta(Y_i = j), \ i \in \mathbb{N}_0, j \in E, \\
p^{(i)} &:= (p_1^{(i)}, \ldots, p_m^{(i)})^t \ , \ i \in \mathbb{N},
\end{aligned}
$$

one obtains for $i > 0$ and $j \leq m$

$$
p_j^{(i)} = \sum_{k \in E} P_\vartheta(Y_i = j \mid Y_{i-1} = k) \, P_\vartheta(Y_{i-1} = k) = \sum_{k=1}^{m} q_{kj} \, p_k^{(i-1)},
$$

i.e. $p^{(i)} = Q^t \, p^{(i-1)}$ and therefore

$$
p^{(i)} = (Q^t)^i \, p^{(0)}.
$$

For the expectation one is interested in (the existence follows from (3.15)) follows

(3.29) Theorem

Let $\sum_{i=1}^{m} p_i^{(0)} = 1$ (i.e. $N \geq 1$ $P_\vartheta - a.s.$),

$$
\begin{aligned}
V^{(1)} &:= (W(1), \ldots, W(m))^t \\
V^{(2)} &:= (V_1^{(2)}, \ldots, V_m^{(2)})^t
\end{aligned}
$$

where $V_i^{(2)} := \sum_{j>m} q_{ij} \, W(j)$, $1 \leq i \leq m$, and

$$
V := V^{(1)} + V^{(2)}.
$$

Then

$$
E_\vartheta\left(\sum_{i=0}^{N} W(Y_i)\right) = \sum_{i=0}^{\infty} \langle V, p^{(i)} \rangle.
$$

Proof: In

$$
E_\vartheta\left(\sum_{i=0}^{N} W(Y_i)\right) = E_\vartheta\left(\sum_{i=0}^{N-1} W(Y_i)\right) + E_\vartheta(W(Y_N)),
$$

one obtains for the two different terms[32]

$$
E_\vartheta\left(\sum_{i=0}^{N-1} W(Y_i)\right) = \sum_{n=1}^{\infty} E_\vartheta\left(\sum_{i=0}^{n-1} W(Y_i) \mid N = n\right) P_\vartheta(N = n)
$$

---

[32] The existence of $E(\sum_{i=1}^{N} | W(Y_i) |)$ (see (3.15)) ensures that the order of summation has no influence.

$$= \sum_{n=1}^{\infty} \sum_{i=0}^{n-1} \sum_{j=1}^{m} W(j) P_\vartheta(Y_i = j \mid N = n) \, P_\vartheta(N = n), \text{ since } i < n$$

$$= \sum_{i=0}^{\infty} \sum_{j=1}^{m} W(j) \sum_{n=i+1}^{\infty} P_\vartheta(Y_i = j \mid N = n) P_\vartheta(N = n)$$

$$= \sum_{i=0}^{\infty} \sum_{j=1}^{m} W(j) p_j^{(i)} = \sum_{i=0}^{\infty} \langle V^{(1)}, p^{(i)} \rangle,$$

$$E_\vartheta(W(Y_N)) = \sum_{j > m} W(j) \, P_\vartheta(Y_N = j)$$

$$= \sum_{j > m} W(j) \sum_{n=0}^{\infty} \sum_{i=1}^{m} P_\vartheta(Y_{n+1} = j \mid Y_n = i) \, P_\vartheta(Y_n = i)$$

$$= \sum_{n=0}^{\infty} \sum_{i=1}^{m} \sum_{j > m} W(j) \, q_{ij} \, p_i^{(n)}$$

$$= \sum_{n=0}^{\infty} \sum_{i=1}^{m} V_i^{(2)} \, p_i^{(n)} = \sum_{n=0}^{\infty} \langle V^{(2)}, p^{(n)} \rangle,$$

i.e. altogether the assertion. $\qquad\qquad\square$

Therefore, the expectations one is interested in can be (numerically) computed by approximating the (infinite) series by finite sums and stopping if e.g. $\sum_{j=1}^{m} p_j^{(i)} < \delta$ or another criterion is fulfilled.

The connection with the LE-method can be perceived by means of the following representation:

$$\begin{aligned} \sum_{n=0}^{\infty} \langle V, p^{(n)} \rangle &= \langle V, \sum_{n=0}^{\infty} p^{(n)} \rangle = \langle V, \sum_{n=0}^{\infty} (Q^t)^n \, p^{(0)} \rangle \\ &= \langle V, (I - Q^t)^{-1} \, p^{(0)} \rangle, \text{ since } \rho(Q^t) = \rho(Q) < 1 \\ &= \langle (I - Q)^{-1} V, p^{(0)} \rangle. \end{aligned}$$

To compute $y = (I - Q)^{-1} V$ one has to solve the system $(I - Q)y = V$; of linear equations; this just leads to the LE-method. Moreover, the identity

$$\sum_{n=0}^{\infty} \langle V, p^{(n)} \rangle = \langle V, \sum_{n=0}^{\infty} (Q^t)^n \, p^{(0)} \rangle$$

shows that the rate of convergence is at least geometrical and is determined by $\rho(Q)$ (since this holds true for $\sum_{n=0}^{\infty} (Q^t)^n$).

Error bounds which are better suited for numerical purposes are obtained by

86

(3.30) Lemma [33]

$$(i)\ \left|\sum_{n=k}^{\infty}\langle V, p^{(n)}\rangle\right| \leq \|p^{(k)}\|_2 \cdot \|V\|_2 / \|I - Q^t\|_2$$

$$(ii)\ \left|\sum_{n=k}^{\infty}\langle V, p^{(n)}\rangle\right| \leq \|Q^t\|_2^k \cdot \|p^{(0)}\|_2 \cdot \|V\|_2 / \|I - Q^t\|_2.$$

Proof: According to Cauchy-Schwarz's inequality

$$\left|\sum_{n=k}^{\infty}\langle V, p^{(n)}\rangle\right| = \left|\langle V, \sum_{n=k}^{\infty} p^{(n)}\rangle\right| \leq \|V\|_2 \cdot \left\|\sum_{n=k}^{\infty} p^{(n)}\right\|_2,$$

where on the one hand

$$\left\|\sum_{n=k}^{\infty} p^{(n)}\right\|_2 = \left\|\sum_{n=0}^{\infty}(Q^t)^n\, p^{(k)}\right\|_2 = \|(I - Q^t)^{-1} p^{(k)}\|_2$$

$$\leq \|(I - Q^t)^{-1}\|_2 \cdot \|p^{(k)}\|_2 \leq \|p^{(k)}\|_2 / \|I - Q^t\|_2$$

(yielding (i)) and on the other hand

$$\|p^{(k)}\|_2 = \|(Q^t)^k p^{(0)}\|_2 \leq \|Q^t\|_2^k \cdot \|p^{(0)}\|_2$$

which proves assertion (ii). $\qquad\square$

For SPPRT's the following cases are of special interest:

(3.31) Corollary

(i)
$$ASC(\vartheta) = \sum_{n=0}^{\infty}\langle C, p^{(n)}\rangle$$

where $C := (c(t(1)), \ldots, c(t(m)))^t$.

(ii)
$$PF(\vartheta) = \sum_{n=0}^{\infty}\langle q_1, p^{(n)}\rangle$$

where $q_1 := (q_{1,m+1}, \ldots, q_{m,m+1})^t$.

(iii)
$$OC(\vartheta) = \sum_{n=0}^{\infty}\langle q_2, p^{(n)}\rangle$$

where $q_2 := (q_{1,m+2}, \ldots, q_{m,m+2})^t$.

(iv)
$$P_\vartheta(N > n) = \langle e, p^{(n)}\rangle$$

where $e = (1, \ldots, 1)^t$.

---

[33] For $x \in \mathbb{C}^m, p \in (0, \infty)$ we denote (as usual)

$$\|x\|_p := \left(\sum_{i=1}^{m} |x_i|^p\right)^{1/p}.$$

Proof: (i) – (iii) follow from (3.29); moreover

$$\langle (1,\ldots,1)^t,\, p^{(n)}\rangle = \sum_{j=1}^{m} P_\vartheta(Y_n = j) = P_\vartheta(N > n). \qquad \square$$

The actual computation of the characteristics is performed in the following way: Let for $n \in \mathbb{N}$

$$PF_n := \sum_{k=0}^{n}\langle q_1, p^{(k)}\rangle, \quad OC_n := \sum_{k=0}^{n}\langle q_2, p^{(k)}\rangle, \quad ASC_n := \sum_{k=0}^{n}\langle C, p^{(k)}\rangle.$$

The values $p^{(n)}$ are computed recursively and then, using PF_{n-1}, OC_{n-1} and ASC_{n-1}, the values PF_n, OC_n and ASC_n. For given $\delta > 0$ one will stop the procedure as soon as[34]

$$|\, 1 - PF_n - OC_n \,| < \delta$$

(according to the fact that

$$PF_n + OC_n + P_\vartheta(N > n) = 1 \qquad \forall n \in \mathbb{N}).$$

If the procedure stops after the n_0-th iteration then

$$P_\vartheta(N > n_0) < \delta,$$

i.e. one obtains lower and upper bounds for the approximations which ensure that the error for $OC(\vartheta)$ as well as for $PF(\vartheta)$ is $< \delta$. Moreover,

$$ASC_{n_0} \leq ASC(\vartheta),$$

i.e. one obtains (unfortunately only) a lower bound for the exact value.

With slight modifications this method can also be used for the stopping rules N_M – but one has, in particular, to take care that the case $N_M = M$ is treated correctly (or M is choosen large enough to neglect this effect).

Similarly to the ILF-method an advantage of the BF method consists in the possibility to utilize the special structure of the matrix Q (as mentioned in (3.18)): On the one hand Q may be stored in an efficient way, on the other hand the algorithm from (3.19)b) can be used for the recursive computation of $p^{(n)}$. To compute one value of the PF-, the OC- and the ASC-function then about $m^2/\,|\,g_1\,|$ operations for $p^{(n)}$ and $4m$ operations for the scalar products are needed per iteration step. Computing only a value of the OC-function (or of the PF- or of the ASC-function) requires nearly the same expenditure as determining all three values since the matrix multiplication needed for $p^{(n)}$, which yields the "leading" term $m^2/\,|\,g_1\,|$, has to be performed in each iteration step.

---

[34]For an actual implementation one should use the stopping criterion

$$|\, 1 - PF_n - OC_n \,| < \delta \quad OR \quad P_\vartheta(N > n) < \delta$$

in order to avoid that an underflow influences the termination of the program. For this reason, one then has to compute $P_\vartheta(N > n)$ (according to (3.31)(iv)).

(3.32) Remark

The basic method of the BF-algorithm may also be used if the transition probabilities depend on the respective number of sub-samples (e.g. if the "continuation function" $\hat{t}$ of (3.5) also depends on the stage i) and therefore the Markov chain $(Z_{\tau_1+\ldots+\tau_n})_{n\in\mathbb{N}}$ is no longer homogeneous. In this case one has to consider a corresponding modification of the recursion for $p^{(n)}$. This means that this modified BF-method can be used to compute characteristics of sequentially planned tests where the size of the next sub-sample depends not only on the present "state" but also on the number of sub-samples taken up to that time. Moreover, if additionally also the "weight-function" W depends on i the BF-algorithm can be modified in an appropriate way (in (3.29) V has to be replaced by V_i). In particular, one can use this method to compute the expected sampling costs of sequentially planned tests if the assumption of stage-independent costs is violated (see (1.23)d)). Therefore, the general method underlying the BF-algorithm is by no means restricted to homogeneous Markov chains with stationary "weight-functions". $\quad\square$

Using the BF-method it is extremely difficult to compute the variance of $ASC(\vartheta)$; on the other hand it is quite easy to determine the expectation and the variance of N since $P_\vartheta(N > n)$ is known (according to (3.31)(iv)).

A disadvantage of the BF-method is that sometimes a large number of iterations is needed to fulfill prescribed accuracy requirements. But for SPPRT's planned in a reasonable way absorption will occur in considerably fewer steps than it does for SPRT's – so far the BF-method will turn out to be more efficient for SPPRT's than for SPRT's (see also § 6).

(III) The EV-method

An obvious idea to overcome this disadvantage is to estimate the "tails" of the series in order to get reasonable results with less expense. Woodall and Reynolds [W/R] took up this idea. It is based on eigenvalues of non-negative matrices; therefore we will call it EV-method.

Q is assumed to be irreducible, i.e. there exists no permutation of the states which transforms Q into

$$\begin{pmatrix} D_1 & \star \\ 0 & D_2 \end{pmatrix}$$

where D_1, D_2 are quadratic matrices – according to [W/R] this means no severe restriction. Then $\lambda := \rho(Q)$ is an eigenvalue of Q (see e.g. [C/M], p. 120) and there exist positive vectors x, y such that

$$Qx = \lambda x \quad , \quad Q^t y = \lambda y;$$

x, y are assumed to be normalized s.t. $\langle x, y \rangle = 1$. Let furthermore

$$k_0 := \#\{\mu \in C : \mu \text{ is eigenvalue of } Q, \ |\mu| = \lambda\}.$$

Then

$$\lim_{n\to\infty} \frac{1}{k_0} \sum_{i=0}^{k_0-1} \left(\frac{Q}{\lambda}\right)^{n+i} = xy^t.$$

(see e.g. [C/M], p. 122). This is equivalent to

$$(\frac{Q}{\lambda})^n \xrightarrow{n\to\infty} xy^t C^{-1},$$

where

$$C = \frac{1}{k_0} \sum_{i=0}^{k_0-1} (\frac{Q}{\lambda})^i.$$

If Q is primitive, i.e. $k_0 = 1$, then

$$(Q/\lambda)^n \xrightarrow{n\to\infty} xy^t.$$

These asymptotic properties of Q^n can be utilized to get asymptotic assertions on $p^{(n)}$ (see also [Se], p. 98):

(3.33) Lemma

 Let Q be irreducible; then

$$\lim_{n\to\infty} p^{(n)}/\|p^{(n)}\| = y/\|y\|.$$

Proof: From

$$p^{(n)} = (Q^t)^n\, p^{(0)} = \lambda^n((\frac{Q}{\lambda})^n)^t\, p^{(0)}$$

follows

$$\frac{p^{(n)}}{\|p^n\|} = \frac{((\frac{Q}{\lambda})^n)^t\, p^{(0)}}{\|((\frac{Q}{\lambda})^n)^t p^{(0)}\|} \xrightarrow{n\to\infty} \frac{(xy^t C^{-1})^t\, p^{(0)}}{\|(xy^t C^{-1})^t\, p^{(0)}\|}.$$

Moreover

$$C^t y = \frac{1}{k_0} \sum_{i=0}^{k_0-1} (\frac{Q^t}{\lambda})^i y = y$$

and therefore $y = (C^t)^{-1} y$. Altogether follows

$$\lim_{n\to\infty} \frac{p^{(n)}}{\|p^{(n)}\|} = \frac{(C^{-1})^t y x^t\, p^{(0)}}{\|(C^{-1})^t y x^t\, p^{(0)}\|} = \frac{y x^t\, p^{(0)}}{\|y x^t\, p^{(0)}\|} = \frac{y}{\|y\|},$$

since $x^t\, p^{(0)} > 0$. $\qquad\square$

This lemma implies that $p^{(n)}$ is, for large n, approximately an eigenvector of Q^t with respect to the eigenvalue λ, i.e.

$$(\star) \qquad p^{(n+i)} = (Q^t)^i\, p^{(n)} \approx \lambda^i\, p^{(n)}.$$

According to (3.29) the characteristics one is interested in allow a representation

$$\sum_{i=0}^{\infty} \langle V, p^{(i)} \rangle.$$

The basic idea is to start with several iterations according to the BF-method and then to utilize $(\star)$. One obtains

$$\sum_{i=0}^{\infty} \langle V, p^{(i)} \rangle = \sum_{i=0}^{n_0-1} \langle V, p^{(i)} \rangle + \sum_{i=n_0}^{\infty} \langle V, p^{(i)} \rangle,$$

90

where the second term is, due to $(\star)$, approximately

$$\langle V, p^{(n_0)}\rangle \sum_{i=0}^{\infty} \lambda^i.$$

This yields altogether

$$\sum_{i=0}^{\infty}\langle V, p^{(i)}\rangle \approx \sum_{i=0}^{n_0-1}\langle V, p^{(i)}\rangle + \langle V, p^{(n_0)}\rangle \frac{1}{1-\lambda}.$$

To make use of this idea, one needs information concerning λ. For this goal one can either use standard algorithms which are suited also for sparse matrices, or utilize $(\star)$: Since $p^{(n)} \approx \lambda p^{(n-1)}$ and therefore

$$\|p^{(n)}\| \approx \lambda\|p^{(n-1)}\|$$

λ may be approximated by $\|p^{(n)}\|/\|p^{(n-1)}\|$.

An appropriate choice of the norm $\|\cdot\|$ is $\|\cdot\|_1$ since these values have to be computed anyway (this kind of determining the "largest" eigenvalue just leads to the power method of numerical mathematics; for primitive Q this procedure always converges).

Applying the EV-method, one will at first carry out several iterations in order to approximate $p^{(n+i)}$ on the basis of $(\star)$, then further iteration steps will be made until the accuracy of approximating λ by $\|p^{(n)}\|_1/\|p^{(n-1)}\|_1$ is sufficiently high. In order to judge the quality of the approximation it seems to be reasonable to use, analogously to the BF-method, the values computed for the OC- and PF-function.

Compared with the BF-method the expense required by the EV-method is nearly the same, but in many cases one obtains results of comparable accuracy (see § 6) considerably faster. The price one has to pay for this advantage is that no exact error bounds (as for the BF-method) are available for the EV-method. The other advantages/disadvantages of the BF-method carry over to the EV-method. For practical purposes one should combine the EV- and BF-method and compute the characteristics using the BF-method if this promises (e.g. for non-primitive matrices Q) shorter execution times.

§ 5 Remarks on the implementation of the algorithms; Examples

The algorithms of § 4 were implemented for some special classes of distributions (binomial-, Poisson- and negative-binomial-distributions) and SPPRT's of the "onion-skin"-type (see Ch. V), i.e. there exist intervals $(k_1, k_2) =: I_1 \supset I_2 \supset \dots$ such that $\hat{t}(x) = j$ for $x \in I_j - \cup_{i>j} I_i$. The program, written in MODULA, was separated into modules for special tasks:

(i) The module <u>GlobalDefinitions</u> contains global definitions and types and, additionally, a procedure which determines the internal representation of SPPRT's in which the number of additional observations is directly assigned to the states.

(ii) The module <u>Densities</u> provides, for the considered distributions (binomial-, Poisson and negative-binomial- distributions) the corresponding probabilities and distribution functions and, moreover, procedures for computing $\gamma_0(\vartheta_1, \vartheta_2)$ and $\gamma_1(\vartheta_1, \vartheta_2)$.

(iii) The module <u>Methods</u> contains the procedures EV, BF and ILE which implement the corresponding methods. For the ILE-method one has, moreover, the choice between "with" and "without" new initialization (the procedures require a representation of the test according to (i)). Additionally there exists an input-parameter "start" which allows us to choose the initial state of the Markov chain for which the values have to be computed. This module contains a local module "Markov" summarizing all operations (in the special implementation of § 4) connected with the matrix Q.

(iv) The module <u>IOManager</u> provides a procedure (the main program) organizing the input/output operations and making some options available.

Furthermore, the module <u>DiophantineApproximation</u> is available which computes an appropriate rational approximation of a real number (see Appendix B).

The program requires the following inputs: Firstly, the sampling costs have to be specified. Then the distribution ($\mathcal{B}(1;p)$-(binomial-), $\mathcal{P}(p)$-(Poisson-) or $Nb(1;p)$-(negative binomial-) distribution) has to be chosen. Thirdly, the parameter values describing the simple hypotheses $H_i(= \{\vartheta_i\})$ have to be fixed. Furthermore the SPPRT in its "onion-skin" representation (with respect to the sequence $\gamma_1(\vartheta_1, \vartheta_2) \sum_{i=1}^{n} X_i - n \, \gamma_0 \, (\vartheta_1, \vartheta_2)$) has to be specified (this simplification was used to get simple inputs). Then one has to choose a method (EV, BF ILE $\sim$ ILE-method *with* or ILO $\sim$ ILE-method *without* new initialization) for the actual computations. Finally a list of those parameter values is needed for which the characteristics have to be evaluated; this is done by giving a lower and an upper value and the step width.

Moreover, one has some additional options: The accuracy ε (standard $= 10^{-4}$) of the diophantine approximation and the accuracy δ (standard $= 10^{-5}$) of the computation of the characteristics can be changed; one may decide to compute, using the ILE-method, also the variance of the ASC-function. Additionally, one has the possibility to determine, for given error bounds, the outer boundaries of an SPPRT according to Wald's approximations (3.7).

The first part of the output of the program consists of a documentation of the input. Then the values $\gamma_0(\vartheta_1, \vartheta_2)$, $\gamma_1(\vartheta_1, \vartheta_2)$, the diophantine approximation and the dimension of Q are given (it is also possible to check these values at an earlier stage of the program to get an impression of the effort needed for the evaluation). The computed values are listed in a table.

At the ILE-method "iASC" and "iOC" denote the number of iterations needed for computing $ASC(\vartheta)$ and $OC(\vartheta)$ resp. Analogously for the BF- and the EV-method "Num" denotes the number of iterations and "Rest" the neglected probability mass i.e. $P_\vartheta(N > Num)$. In the EV-method the sign "+" behind the value of "Num" indicates that all characteristics are computed using only the EV-method; if this sign is missing, some results are obtained using the BF-method (this indicates that the matrix Q fails to fulfill all conditions (see § 4) which are sufficient for quick convergence of the EV-method). Finally the execution times are given.

Listings of the modules with additional comments are given in Appendix B.

In order to illustrate the properties of SPPRT's and of the different algorithms, we consider some examples:

(3.34) Example
Coming back to the example on p. 10, we consider the case of $\mathcal{B}(1,p)$ (binomial-) distributions with

$$p_1 = 0.90016837105, \quad p_2 = 0.95.$$

a) Since the SPRT's are special cases of SPPRT's, we have the opportunity to check the results delivered by our algorithms. The results are exactly the same as on pp. 14-16; the OC-function of the SPRT $\delta_{1/9,9}$ is shown in figures 3.2 and 3.3 resp., some values are given in table 3.1. For the sampling cost function given by

$$(\star) \qquad c(\vartheta, n) = c_0 + n$$

we obtain

  (i) for $c_0 = 0$ just the ASN-function; this is shown in figures 3.4 and 3.5 resp., some special values are given in table 3.2,

  (ii) for $c_0 = 1$ the ASC-function shown in figures 3.6/3.7; some special values are given in table 3.3,

  (iii) for $c_0 = 20$ the corresponding curves/values are given in figures 3.8/3.9 and table 3.4 resp.

p	$SPRT\ \delta_{1/9,9}$	$SPPRT\ b)$	$SPPRT\ c)$
0.8	0.999808	0.999821	0.999968
0.82	0.999395	0.999444	0.999829
0.84	0.998001	0.998185	0.999126
0.86	0.993062	0.993622	0.995729
0.88	0.974482	0.975745	0.979969
0.9	0.901281	0.902955	0.909829
0.92	0.646052	0.647749	0.653595
0.94	0.211822	0.214359	0.215072
0.96	0.0222622	0.0218219	0.0210901
0.98	0.000544063	0.000388764	0.000230046
1.0	$1.0234 \cdot 10^{-24}$	$2.50823 \cdot 10^{-11}$	$1.0087 \cdot 10^{-10}$

<u>Table 3.1:</u>
Values of the OC-function of the SPRT $\delta_{1/9,9}$
and of two special SPPRT's

p	$SPRT\ \delta_{1/9,9}$	$SPPRT\ b)$	$SPPRT\ c)$
0.8	25.9784	44.3679	80.2527
0.82	30.7198	47.6214	80.8359
0.84	37.5494	53.074	82.5344
0.86	48.0799	62.3227	87.0662
0.88	65.6326	78.4854	98.1866
0.9	96.135	106.848	122.411
0.92	137.173	143.842	157.272
0.94	135.107	138.55	151.705
0.96	88.0161	91.3527	107.531
0.98	56.8295	59.1346	83.4952
1.0	41.0	40.0	80.0

<u>Table 3.2</u>
Values of the ASN/ASC-function of the
SPRT $\delta_{1/9,9}$ and two special SPPRT's for $c_0 = 0$

p	$SPRT\ \delta_{1/9,9}$	$SPPRT\ b)$	$SPPRT\ c)$
0.8	51.9568	45.5513	81.2582
0.82	61.4397	48.9245	81.8534
0.84	75.0988	54.5645	83.5856
0.86	96.1598	64.1107	88.204
0.88	131.265	80.7673	99.5283
0.9	192.27	109.983	124.181
0.92	274.346	148.112	159.643
0.94	270.213	142.73	153.987
0.96	176.032	94.1591	109.058
0.98	113.659	60.9196	84.5751
1.0	82.0	41.0	81.0

<u>Table 3.3</u>
Values of the ASN/ASC-function of the
SPRT $\delta_{1/9,9}$ and two special SPPRT's for $c_0 = 1$

94

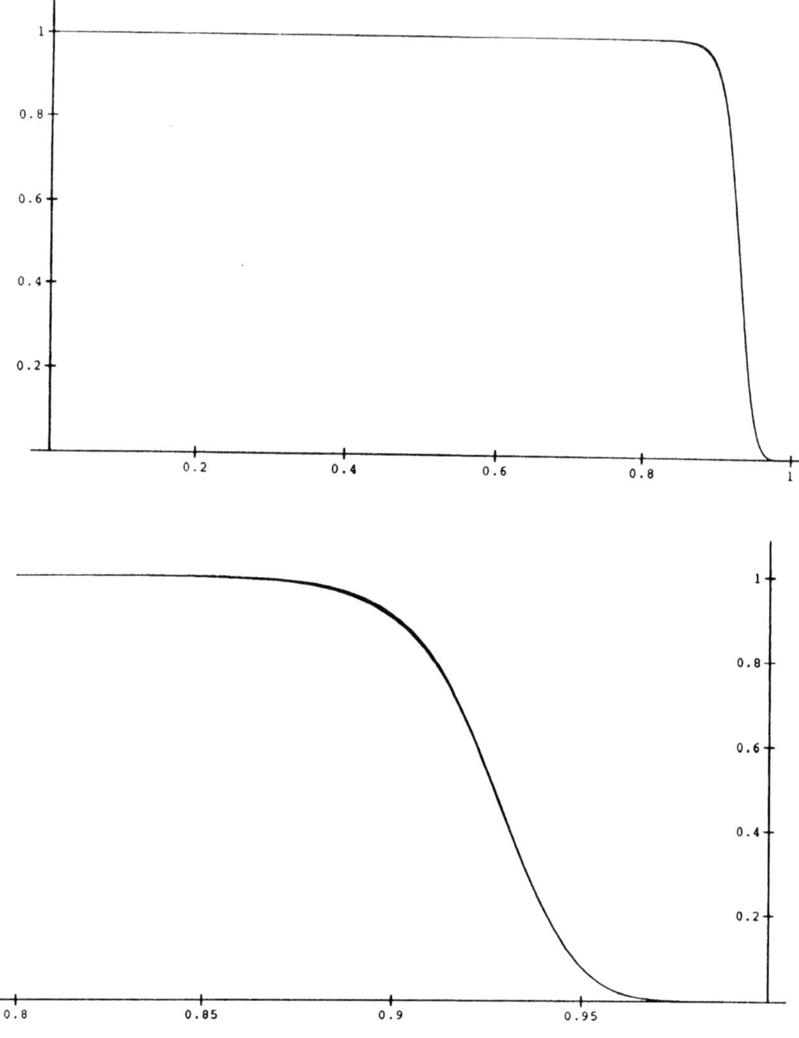

<u>Figures 3.2/3.3:</u>
OC-functions of the SPRT $\delta_{1/9,9}$
and of two special SPPRT's

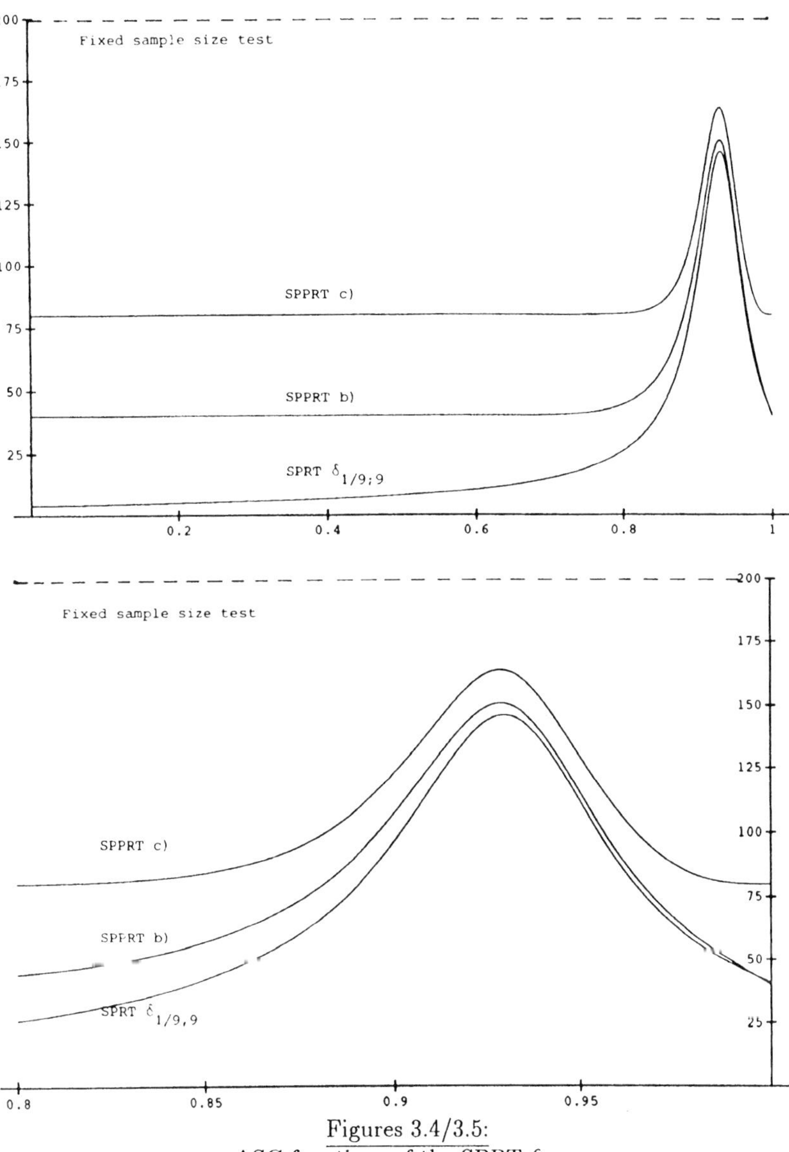

Figures 3.4/3.5:
ASC-functions of the SPRT $\delta_{1/9,9}$
and two special SPPRT's for $c_0 = 0$

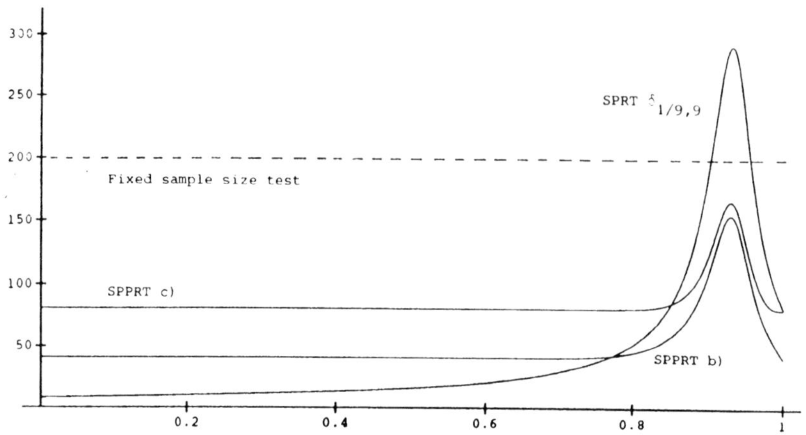

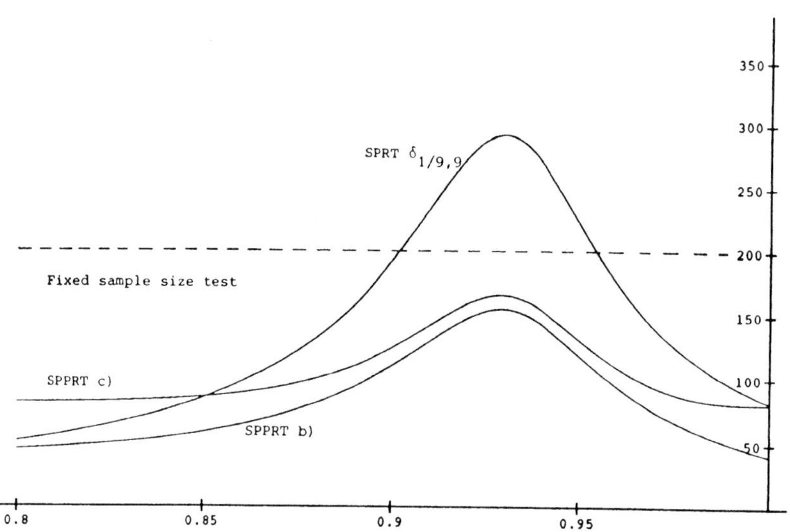

Figures 3.6/3.7

ASC-functions of the SPRT $\delta_{1/9,9}$
and two special SPPRT's for $c_0 = 1$

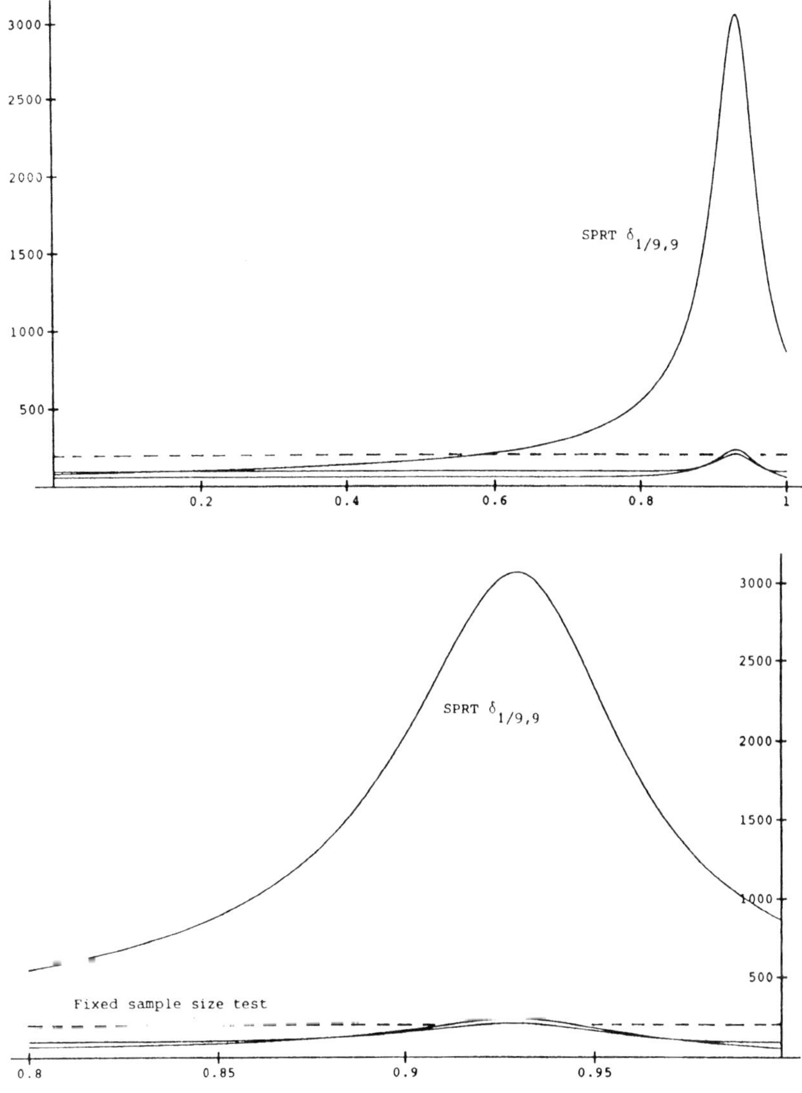

Figures 3.8/3.9:
ASC-functions of the SPRT $\delta_{1/9,9}$
and two special SPPRT's for $c_0 = 20$

p	$SPRT\ \delta_{1/9,9}$	$SPPRT\ b)$	$SPPRT\ c)$
0.8	545.546	68.0368	100.362
0.82	645.117	73.6831	101.186
0.84	788.537	82.8846	103.559
0.86	1009.68	98.083	109.821
0.88	1378.28	124.124	125.021
0.9	2018.83	169.537	157.815
0.92	2880.64	229.233	204.706
0.94	2837.24	222.162	197.346
0.96	1848.34	147.481	138.076
0.98	1193.42	94.8348	105.093
1.0	861.0	60.0	100.0

<u>Table 3.4</u>
Values of the ASN/ASC-function of the
SPRT $\delta_{1/9,9}$ and two special SPPRT's for $c_0 = 20$

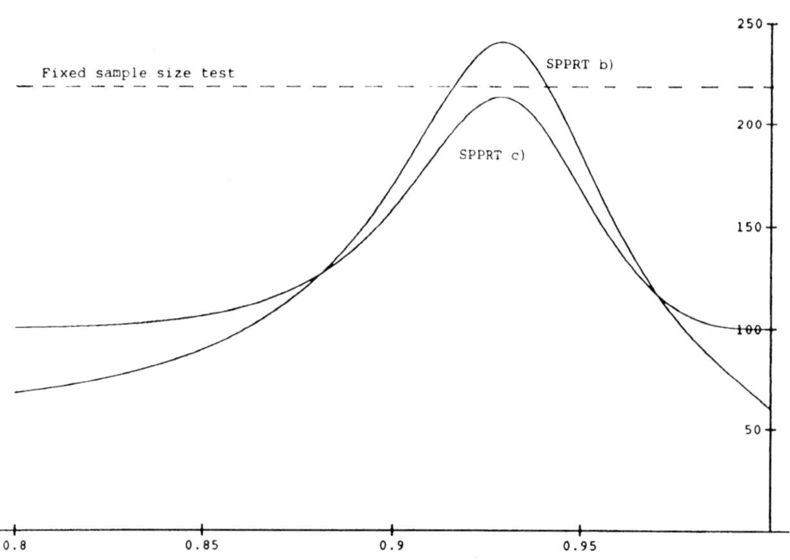

<u>Figure 3.10:</u>
ASC-functions of the SPPRT's b) and c)
for $c_0 = 20$

b) To illustrate the effects of grouped sampling we next consider the SPPRT of "onion skin" type given by the intervals

$$I_j = \left\{ \begin{array}{l} (-1.73, 1.75) \text{ for } j = 1, \ldots, 10; \\ (-1.5, 1.5) \text{ for } j = 11, \ldots, 20; \\ (-1.0, 1.0) \text{ for } j = 21, \ldots, 40 \end{array} \right.$$

(i.e. only the sub-sample sizes 10, 20 and 40 are possible). Here we obtain nearly the same OC-function as for the SPRT $\delta_{1/9,9}$; see figures 3.2 and 3.3 and table 3.1. For the sampling cost function $(\star)$ one gets

(i) for $c_0 = 0$ the ASC-function shown in figures 3.4/3.5 and table 3.2 (obviously in this case the SPRT $\delta_{1/9,9}$ is "better"),

(ii) for $c_0 = 1$ the ASC-function shown in figures 3.6/3.7 and table 3.3; here the SPPRT turns out to be highly superior to the SPRT $\delta_{1/9,9}$ in the most interesting interval $[0.8, 1.0]$,

(iii) for $c_0 = 20$ the ASC-function shown in figures 3.8-3.10 and table 3.4 resp.; here the advantages of grouped sampling become even more evident (the behaviour of the SPRT $\delta_{1/9,9}$ is desasterously bad).

c) To illustrate the intuitively obvious effect that for higher fixed costs the sizes of subsamples should increase we furthermore consider the SPPRT given by the intervals

$$I_j = \left\{ \begin{array}{l} (-1.5, 1.33) \text{ for } j = 1, \ldots, 30, \\ (-0.7, 0.7) \text{ for } j = 31, \ldots, 80. \end{array} \right.$$

Again this test has nearly the same OC-function as the previous ones (see figures 3.2/3.3 and table 3.1). For the values

$$c_0 = 0, \; 1 \; \text{ and } \; 20$$

the ASC-function is shown in the figures 3.4 - 3.10 and tables 3.2 - 3.4. A comparison of the two SPPRT's is given in figure 3.10.

These values show, moreover, that the SPPRT's of part b) and c) are reasonable candidates for *monitoring strategies* for clinical trials: Firstly, an initial investigation with 40 patients (in case b); 80 in case c)) is made. If no dramatic effects occur another group of 40 (or 80 resp.) patients is treated; but if either a clear decision is possible or an indication for a soon decision is obtained, one switches over to no further treatment or to a smaller group size (of sizes 10 or 20 in case b) and 30 in case c)). Our algorithms allow to compute (numerically) the exact OC-function as well as the expected sampling costs needed for this clinical trials with interim monitoring (for this aspect see also the review article by Jennison and Turnbull [J/T]).

A graphical illustration of our three special sampling designs is given in figure 3.11.

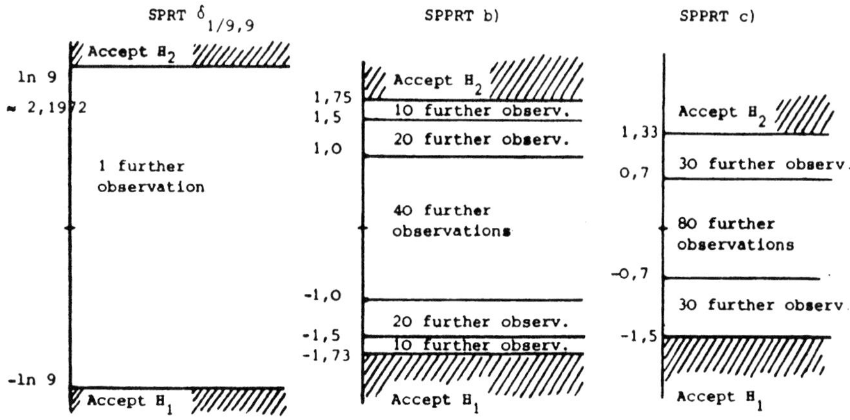

Figure 3.11 Graphical representation of the SPRT $\delta_{1/9,9}$
and two special SPPRT's

To get an impression of the execution times needed by the different methods, the times to compute $OC(\vartheta)$ and $E_\vartheta g(\tau)$ at the parameter values

$$\vartheta = 0.80 + i \cdot 0.02 \quad (i = 0, \ldots, 10)$$

for the three tests mentioned above were registered for our algorithms BF, EV, ILE and ILO. The corresponding values, given in table 3.5, give the sum of "user" and "sys"-times (in seconds) found out by the time-command under UNIX on a SUN 3; these values allow a realistic comparison of the different execution times. But one should notice that all methods are iterative, for which no exact proposition on the accuracy of the solutions is known; hence smaller execution times of a special algorithm might on the one hand be explained by a higher "speed" of this method but on the other hand also by less accurate results.

	BF	EV	ILE	ILO
SPRT$\delta_{1/9,9}$	490.3	389.3	818.9	708.4
SPPRT b)	19.3	16.5	22.7	22.6
SPPRT c)	13.9	12.9	17.2	17.2

Table 3.5
Execution times for computing the characteristics
$OC(\vartheta)$ and $E_\vartheta g(\tau)$ at $\vartheta = 0.8 + i \cdot 0.02 \quad (i = 0, \ldots, 10)$
(measured under UNIX on a SUN 3; user + sys-time)

Nevertheless the comparison of these values yields some interesting results. Firstly, the properties to be expected due to the construction are confirmed by these data:

The EV-method turns out to be quicker than the BF-method, and the ILO-method is quicker than the ILE-method. Moreover, in these examples the BF/EV-method is always superior to the ILE/ILO-method. But also examples with reversed preference may be constructed (see [Me]); hence we conjecture that the execution times needed by the BF/EV- and the ILE/ILO-method are essentially of the same order of magnitude.

Additionally, one should notice the "smoothing" effect which occurs for the SPPRT's in comparison to the SPRT: Whilst the differences at the SPRT $\delta_{1/9,9}$ are significant, they are considerably smaller for the SPPRT's b) and c). An obvious explanation of this effect is the more "favourable" form of the matrix Q at SPPRT's – for relatively large sub-sample sizes the probabilities of absorption are much higher than for "one at a time" sampling. This is also the reason why the computation of the characteristics of SPPRT's needs considerably less time than the corresponding computation for the SPRT (although the problem itself is quite similar).

Hence for computing the characteristics of SPRT's one should use the algorithms constructed for these special purpose (see [Eg], [E/M]); but for handling "reasonable" SPPRT's (i.e. procedures with larger sub-sample sizes) the presented methods are very well suited.

§ 6   Remarks on the comparison of the methods and on convergence-improvements for the BF-/EV-method

It was already mentioned that the pure LE-method is of very restricted interest since the special structure of the matrix Q cannot be used in a satisfactory way.

To compare the BF- and the EV-method – in both cases the speed of convergence is, according to (3.29), determined by $\rho(Q)$ – is relatively simple since the EV-method always requires a smaller number of iterations and this with nearly the same computational effort. But if a given error bound for the PF- or OC-function has to be guaranteed one will use the BF-method. In a test-study of Meyerthole [Me] the EV-method in many cases turned out to be much better than the BF-method.

A general comparison of the ILE-method without (ILO) and with new initialization (ILE) seems to be impossible. In the test-study [Me] sometimes only minimal, in other cases a considerable saving of execution time was observed.

A (theoretical) comparison between BF- and EV-method on the one side and ILE-method on the other side also seems hardly to be possible: According to (3.25) the speed of convergence of the ILE-method is determined by $\rho(B_1)$. Since $\rho(B_1) \leq \rho(Q)$ (see (3.26)) a natural conjecture is that the ILE-method is the best one. But since the operations of the ILE-method are more time-consuming, the assertions on the number of operations per iteration are concerned with the *order* of magnitude only, and it makes, for the ILE-method, an important difference whether one has to compute the OC-function only or the ASC-function as well, it is essential for a comparison how much smaller $\rho(B_1)$ is than $\rho(Q)$. Therefore *no general preference* with respect to the execution times seems to hold between the ILE- and the BF-/EV-method (in all three cases the demand of store is approximately the same). Indeed, in the test-study of Meyerthole [Me] the ILE-method in some examples was superior and in other cases inferior to the BF-/ EV-method. An important disadvantage of the BF-/EV-method is, of course, that it is not possible to compute the variance of the sampling costs. In this case one will use the ILE-method, which means that also the ASC-function has to be computed using the ILE-method.

Using the BF-/EV-method

$$\sum_{i=0}^{\infty}\langle V, p^{(i)}\rangle = \sum_{i=0}^{\infty}\langle V,(Q^t)^i p^{(0)}\rangle$$

is computed by "summing up" the single terms; the speed of convergence is determined by $\rho(Q)$. An obvious idea for improving the speed of convergence is to combine several summands. Let $g \in \mathrm{IN}$; then

$$\sum_{i=0}^{\infty}\langle V, p^{(i)}\rangle = \sum_{i=0}^{\infty}\sum_{k=0}^{g-1}\langle V,(Q^t)^{gi+k}p^{(0)}\rangle$$

$$= \sum_{i=0}^{\infty}\langle V,(Q^t)^{gi}\sum_{k=0}^{g-1}(Q^t)^k\, p^{(0)}\rangle.$$

This is essentially the same formula as before; Q is replaced by Q^g and $p^{(0)}$ by $\sum_{k=0}^{g-1}(Q^t)^k p^{(0)}$ According to $\rho(Q^g) = (\rho(Q))^g$ and $\rho(Q) < 1$ the speed of convergence of this series is, for $g \geq 2$, higher than that of the original one. Moreover, if the direct BF-method needs n summands here at most $\lceil n/g \rceil$ terms are necessary. Hence this idea seems to be very promising. But the next assertion shows that the gained advantages are (more than) compensated by the fact that the matrix Q^g no longer has the same pleasant structure as Q:

(3.35) Lemma

Let $Q^g = (q_{ij}^{(g)})_{1\leq i,j\leq m}$; then it follows from $q_{ij}^{(g)} \neq 0$, that there exist

$$n_2,\ldots,n_g \in \{1,\ldots,m\}$$

such that

$$j - i + g_0(t(i) + \sum_{k=2}^{g} t(n_k)) \equiv 0 \mod |g_1|.$$

Proof (by induction on g): For $g = 1$ the assertion follows from (3.18). For $g \geq 2$ it follows from

$$q_{ij}^{(g)} = \sum_{k=1}^{m} q_{ik}^{(g-1)}q_{kj} \neq 0,$$

that there exists an $n_g \in \{1,\ldots,m\}$ such that

$$q_{i,n_g}^{(g-1)} \neq 0 \neq q_{n_g,j}.$$

Hence on the one hand one obtains from (3.18)

$$j - n_g + g_0\, t(n_g) \equiv 0 \ \ \mathrm{mod}\ \mid g_1 \mid,$$

and on the other hand it follows from the previous step of induction that there exist $n_2, \ldots, n_{g-1} \in \{1, \ldots, m\}$ s.t.

$$n_g - i + g_0\left(t(i) + \sum_{k=2}^{g-1} t(n_k)\right) \equiv 0 \ \ \mathrm{mod}\mid g_1 \mid.$$

By summation mod $\mid g_1 \mid$ one obtains the assertion. $\qquad\square$

Now let

$$b := \#\{\sum_{k=2}^{g} t(n_k) \ \mathrm{mod}\ \mid g_1 \mid : 1 \leq n_k \leq m\}.$$

Then Q^g has approximately b-times more non-trivial entries than Q (these are, moreover, very irregularly distributed). This means that Q^g is, already for small b and g, no longer sparse and that the pleasant structure of the non-trivial entries (given in (3.18)) is lost.

The price one has to pay for the reduction of the spectral radius is that one has to handle the matrix Q^g, which no longer has the structure given in (3.18) and which does not allow to use the efficient algorithms from (3.19). Moreover, for the construction of Q^g one has to perform matrix multiplications which on the one hand are expensive and on the other hand require to store and administer (at least) 2 matrices. The advantage of a higher speed of convergence will, in the general case of SPPRT's, be more than compensated by these disadvantages.

But in the special case $t(i) \equiv 1 \ \mathrm{mod}\mid g_1\mid\forall i$, which is given in particular for (purely sequential) SPRT's (but which hardly occurs for other tests), the structure of Q is preserved:

(3.36) Corollary

Assume that $t(i) \equiv 1 \ mod \mid g_1 \mid\ \forall i$. Then it follows from $q_{ij}^{(n\mid g_1\mid)} \neq 0$ that

$$j - i \equiv 0 \ \ mod \ \mid g_1 \mid.$$

This corollary says that the sub-Markov chain described by $Q^{\mid g_1\mid}$ may be decomposed into $\mid g_1 \mid$ disjoint classes. This can be used to construct a highly efficient algorithm for SPRT's ([Me], p. 81 ff) which was developed by Eger [Eg] in another way. Unfortunately the method cannot be carried over to SPPRT's since it is based on the following idea: Starting at state s of a Markov chain associated with the SPRT, maybe all states $1, \ldots, m$ can be reached by an appropriate number of steps, but in any case after $\mid g_1 \mid$ steps either absorption has occured or the chain has reached a state in

$$S := \{j \in \{1, \ldots, m\} : j - s \equiv 0 \ \ \mathrm{mod}\mid g_1 \mid\},$$

and this repeats after further $\mid g_1 \mid$ steps. Therefore it is sufficient to consider the states in S (with modified transition probabilities); this yields a considerable smaller transition matrix. Contrary to this situation the Markov chain associated with an SPPRT leads (according to (3.34)) to a much stronger "mixing" of the states: After a few steps already "many" states can be reached, and for a given number of steps there will not exist a "small" subset of states which the Markov chain must reach. But since the method of Eger [Eg] is just based on properties of SPRT's which are no longer valid for SPPRT's it is impossible to generalize his method.

Due to a similar reason also the method of Enkawa/Mori [E/M] cannot be carried over to SPPRT's: This method makes use of the fact that the Markov chain associated with an SPRT is translation invariant (s. p. 72) and that the computation of absorption-probabilities of translation invariant Markov chains leads to a difference equation with constant coefficients (which can be solved or simplified using generating functions). But it is just a characteristic property of SPPRT's that the number of additional observations – and therefore the distribution of the next state – depends on the current state and not only on the difference between the states. Hence one obtains a difference equation with *variable coefficients*. Unfortunately there does not exist any standard solution concept for this kind of difference equations. Moreover, the method of Enkawa/Mori essentially makes use of the fact that Q contains not too many different values; this property is also lost for sequentially planned tests.

Finally we mention an aspect which arises in connection with convergence-improvements for the BF-/EV-method; It might turn out that algorithms for SPPRT's lead to more efficient algorithms for SPRT's. For dichotomous variables e.g. an SPRT for many states of the corresponding Markov chain needs a certain minimal number of steps to reach a boundary (comp. also the curtailed inspection plans (1.7)). From the point of view of an SPPRT these steps may be combined to a sub-sample without changing the error probabilities, and also the ASN-function of the SPRT can be computed by choosing $c(n) = c \cdot n$ as sampling costs. In this way, an absorbing state will be reached by considerably less steps. This leads (using the BF-/EV-method) to a significant saving of execution time (for Bernoulli-SPRT's this was carried out by Meyerthole [Me]).

IV. Bayes-optimal sequentially planned decision procedures

§ 1   Introduction

In Chapter I we defined (see (1.22)/(1.25)) sequentially planned statistical decision procedures; Chapter III contained some examples of such procedures. In the sequel we consider the construction of "optimal" procedures for general sequential statistical decision problems

$$\mathcal{P} = ((\mathcal{X}, \mathcal{B}),\ \Theta,\ (P_\vartheta)_{\vartheta \in \Theta},\ \mathcal{A}, (\mathcal{B}_a)_{a \in A}, (D, \mathcal{D}), (L_a)_{a \in A}, c)$$

(see definition (1.22)). To illustrate the situation we mention two especially simple examples:

(4.1) Example (Acceptance/rejection quality control "with replacement";
   see example (1.1)/(1.14)a))

Batches of size m with unknown probability ϑ for a defective item are to be subjected to an acceptance/rejection quality control. The investigation of n items causes inspection costs

$$c_0 + c \cdot n \qquad \text{(i.e. fixed costs plus linear costs)};$$

rejecting a batch leads to costs K_r (e.g. due to a total inspection), accepting a batch causes costs of $\vartheta \cdot K$ (e.g. due to complaints/guaranties). For a maximal number $k (\leq m)$ of stages this situation may analogously to (1.14)a) be described by

$$
\begin{aligned}
\mathcal{X} \;&=\; \{0,1\}^m = \{(x_1, \ldots, x_m) : x_i \in \{0,1\}, 1 \leq i \leq m\} \\
&\qquad (\text{where } 0 \sim \text{ effective}, 1 \sim \text{ defective}), \\
\mathcal{B} \;&=\; \mathcal{P}(\mathcal{X}); \;\Theta = [0; 1],\; P_\vartheta = (\mathcal{B}(1, \vartheta))^m, \\
A \;&:=\; \{(a_1, \ldots, a_j) : 1 \leq j \leq k,\ 1 \leq a_i \leq m, 1 \leq i \leq j\} \cup \{()\} \\
\mathcal{A} \;&=\; \{a \in A \text{ s.t. } \sum_{i=1}^{j} a_i \leq m\}, \\
\mathcal{B}_a \;&=\; \sigma(\pi_1, \ldots, \pi_{g(a)}) \text{ for } g(a) \leq m \text{ and } \mathcal{B}_a = \mathcal{B} \text{ elsewhere}, \\
D \;&=\; \{\text{acceptance, rejection}\},\ \mathcal{D} = \mathcal{P}(D),
\end{aligned}
$$

$$
L_a(\vartheta, d, x) =
\begin{cases}
K_r & d \sim \text{ rejection} \\
& \quad \text{for} \\
\vartheta \cdot K & d \sim \text{ acceptance}
\end{cases}
$$

and

$$c(\vartheta, n) = c_0 + c \cdot n. \qquad \qquad \Box$$

Here one neglects that during an actual acceptance/rejection quality control in case of rejection the inspected items need no further inspection and that in case of acceptance the detected defective items (in the combined sample) will not be delivered. Taking these aspects into account ("sampling without replacement") we arrive at a more realistic example:

(4.2) Example (acceptance/rejection quality control; attribute sampling)

Batches of size m are subjected to an acceptance/rejection quality control. The inspection of n items causes costs of $c_0 + c \cdot n$; rejecting a batch leads to costs k_r for each item not yet inspected (e.g. inspection costs), if a batch is accepted each (not yet detected) defective item causes costs 1; comp. e.g. Pfanzagl [Pf], Schüler [Schü] or Abel [Ab 1], [Ab 3]. It seems to be obvious to describe this situation as a sequential statistical decision problem in the following way

$$\mathcal{X} = \{0,1\}^m, \ \mathcal{B} = \mathcal{P}(\mathcal{X}), \ \Theta = \{0,\ldots,m\};$$

($\vartheta \in \Theta$ means the unknown number of defective items),

P_ϑ : uniform distribution on $\mathcal{X}_\vartheta := \{(x_1,\ldots,x_m) \in \mathcal{X} : \sum_{i=1}^{m} x_i = \vartheta\}$,

$\mathcal{A}, \mathcal{B}_a, D$, and $\mathcal{D}$ as in (4.1),

$$L_a(\vartheta, d, x) = \begin{cases} k_r(m - g(a)) & d \sim \text{rejection} \\ & \text{for} \\ \vartheta - \sum_{i=1}^{g(a)} x_i & d \sim \text{acceptance} \end{cases}$$

and

$$c(\vartheta, n) = c_0 + c \cdot n. \qquad\qquad \square$$

For the decision problems of example (4.1) and (4.2) the experiences with previous batches will, in many cases, lead to an a priori-knowledge about the parameter ϑ. If this knowledge can be made precise in form of an a priori distribution ξ on Θ – i.e. if the parameter may be interpreted as a random variable with distribution ξ – a statistical decision procedure (τ, φ) leads to expected expenses

$$R(\xi, (\tau, \varphi)) = \int_\Theta \int_\mathcal{X} [\int_D L_{\tau(x)}(\vartheta, e, x)\varphi_{\tau(x)}(x, de)] + C(\vartheta, \tau(x)) \, dP_\vartheta(x) \, d\xi(\vartheta)$$

(where $C(\vartheta, a)$ is defined as in (1.23)d)). Then an obvious goal will consist in constructing a decision procedure which minimizes these expected expenses.

Similarly as in these examples, in many statistical decision problems one has experience/prior knowledge on the parameter ϑ. If these can be specified in the form of an a priori-distribution, such situations may be described in the following way:

(4.3) Definition

    a) A general <u>Bayesian sequential decision problem</u> is a tuple

$$(P, \mathcal{T}, \xi)$$

consisting of

- *a general sequential statistical decision problem P (see (1.22)a),*

- *a σ-algebra $\mathcal{T}$ over the parameter-space Θ such that the mappings*

$$\vartheta \mapsto P_\vartheta(B), \ \vartheta \mapsto C(\vartheta, a)$$

are measurable for each $B \in \mathcal{B}, a \in A$, and such that the mappings

$$(\vartheta, d, x) \mapsto L_a(\vartheta, d, x) \,, a \in A$$

are $(\mathcal{T} \otimes \mathcal{D} \otimes \mathcal{B}_a, \overline{\mathrm{IB}^1})$-measurable,

- *a probability distribution ξ on (Θ, T), the a priori-distribution.*

b) *The Bayes-risk with respect to the a priori-distribution ξ of a sequentially planned statistical decision procedure (τ, φ) is given by*

$$R(\xi, (\tau, \varphi)) \; := \; \int_\Theta \int_{X'} [\int_D L_{\tau(x)}(\vartheta, e, x) \varphi_{\tau(x)}(x, de)]$$
$$+ \, C(\vartheta, \tau(x)) \; dP_\vartheta(x) \; d\xi(\vartheta).$$

c) *A sequentially planned decision procedure $(\tau^\star, \varphi^\star)$ is called Bayes-procedure with respect to ξ if*

$$R(\xi, (\tau^\star, \varphi^\star)) = \inf_{(\tau, \varphi)} R(\xi, (\tau, \varphi)).$$

Judging sequentially planned decision procedures on the basis of their Bayes-risk leads (for Bayesian decision problems) to a linear order on these procedures and therefore to a satisfactory optimality criterion – Bayes-procedures are *optimal* in this sense.

§ 2   Bayes-procedures

As for purely sequential decision problems (1.12), it can be shown also for general Bayesian sequential decision problems that the construction of Bayes-procedures may be split up into two nearly independent steps, namely

- to construct a family $(\varphi_a^\star)_{a \in A}$ of Bayesian terminal decision functions $\varphi_a^\star$ for the non-sequential decision problems with fixed sample sequence a, and

- subsequently to solve a problem of optimal sampling (see Ch. II):

(4.4) Theorem

For a general Bayesian sequential decision problem let $\varphi^\star = (\varphi_a^\star)_{a \in A}$ be a terminal decision procedure such that

$$R(\xi, (a, \varphi^\star)) = \inf_\varphi R(\xi, (a, \varphi)) \qquad \forall a \in A.$$

Then

$$R(\xi, (\tau, \varphi^\star)) = \inf_\varphi R(\xi, (\tau, \varphi))$$

for each sequential sampling plan τ.

A decision procedure $\varphi^\star$ with this property is called *Bayesian terminal decision procedure* (with respect to ξ).

Proof (see [Ir 1], p. 32): Assume there exist a sequential sampling plan τ and a terminal decision procedure φ such that

$$R(\xi, (\tau, \varphi)) = \int_\Theta \sum_{a \in A_{\{\tau = a\}}} \int \, [\int_D L_a(\vartheta, e, x) \varphi_a(x, de)] + C(\vartheta, a) \; dP_\vartheta(x) \; d\xi(\vartheta) < R(\xi, (\tau, \varphi^\star)).$$

108

Then there also exists an $a \in A$ such that

$$\int_\Theta \int_{\{\tau=a\}} \int_D L_a(\vartheta, e, x)\, \varphi_a(x, de)\, dP_\vartheta(x)\, d\xi(\vartheta) <$$
$$< \int_\Theta \int_{\{\tau=a\}} L_a(\vartheta, e, x)\, \varphi_a^\star(x, de)\, dP_\vartheta(x)\, d\xi(\vartheta).$$

Defining

$$\hat{\varphi}_a := \varphi_a\, 1_{\{\tau=a\}} + \varphi_a^\star\, 1_{\{\tau \neq a\}}$$

yields a terminal decision function $\hat{\varphi}_a$ s.t.

$$R(\xi, (a, \hat{\varphi})) < R(\xi, (a, \varphi^\star))$$

contradicting the assumption on $\varphi^\star$. $\qquad\qquad\qquad\qquad\qquad\qquad\qquad\square$

This theorem serves as a key for solving the first part of constructing a Bayes procedure. It means that a Bayesian terminal decision procedure is obtained by solving, for each $a \in A$, a "classical" Bayesian decision problem – and for this aim a well-developed theory is available (see e.g. [Be], Ch. 4).

(4.5) Corollary

Let $\varphi^\star$ be a Bayesian terminal decision procedure and let $\tau^\star$ be a sequential sampling plan such that

$$R(\xi, (\tau^\star, \varphi^\star)) = \inf_\tau R(\xi, (\tau, \varphi^\star));$$

then $(\tau^\star, \varphi^\star)$ is a Bayes-procedure with respect to ξ.

For a family $(P_\vartheta)_{\vartheta \in \Theta}$ of probability distributions and an a priori-distribution ξ

$$W(T \times B) := \int_T P_\vartheta(B)\, d\xi(\vartheta)$$

leads to a probability distribution W on $T \otimes B$; its marginal measure on B, defined by

$$P_\xi(B) := W(\Theta \times B) = \int_\Theta P_\vartheta(B)\, d\xi(\vartheta),$$

will be denoted by P_ξ. For the construction of Bayes procedures with respect to ξ the conditional distributions of ϑ (with respect to W) under $\mathcal{B}_a$ play an important role.

(4.6) Definition

An a posteriori-distribution of ϑ under (the condition) $\mathcal{B}_a$, $a \in A$, for a given a priori-distribution ξ is a stochastic kernel ξ_a from $(\mathcal{X}, \mathcal{B}_a)$ to (Θ, T) such that

$$\int_T P_\vartheta(B)\, d\xi(\vartheta) = \int_B \xi_a(x, T)\, dP_\xi(x) \quad \forall B \in \mathcal{B}_a, T \in \mathcal{T}.$$

On the usual way of algebraic induction one obtains:

(4.7) Remark

If for $a \in A$ there exists an a posteriori-distribution ξ_a of ϑ under $\mathcal{B}_a$ then for each $(\mathcal{T} \otimes \mathcal{B}_a)$-measurable mapping $h : \Theta \times \mathcal{X} \to [0, \infty]$ one obtains

$$\int_\Theta \int_\mathcal{X} h(\vartheta, x) dP_\vartheta(x) \, d\xi(\vartheta) = \int_\mathcal{X} \int_\Theta h(\vartheta, x) \, \xi_a(x, d\vartheta) \, dP_\xi(x).$$

With the help of a posteriori-distributions the second part of the construction of a sequential Bayes-procedure may be reduced to problems of optimal sampling (as treated in Chapter II): Having determined (using (4.4)) a Bayesian terminal decision procedure then one needs (according to (4.5)) a sampling plan τ which minimizes the risk $R(\xi, (\tau, \varphi^\star))$. But if one has, for each $a \in \mathcal{A}$, an a posteriori-distribution ξ_a of ϑ under $\mathcal{B}_a$ this risk may be rewritten in the form

$$R(\xi, (\tau, \varphi^\star)) = \sum_{a \in \mathcal{A}} \int_{\{\tau = a\}} \int_\Theta \int_D \left(\int_D L_a(\vartheta, e, x) \varphi_a^\star(x, de) + C(\vartheta, a) \right) \xi_a(x, d\vartheta) dP_\xi(x).$$

Looking for a τ which *minimizes* this term is therefore equivalent to solving a problem of optimal sampling with the pay-offs

$$Z_a := \begin{cases} -\int_\Theta \left(\int_D L_a(\vartheta, e, x) \varphi_a^\star(x, de) + C(\vartheta, a) \right) \xi_a(x, d\vartheta) & \text{for } a \in \mathcal{A} \\ -\infty & \text{elsewhere.} \end{cases}$$

As known from classical theory, interchanging the integrations over Θ and $\mathcal{X}$ by using a posteriori-distributions in many cases (e.g. finite D) leads to a construction of Bayesian terminal decision procedures by considering the *a posteriori risks*

$$\int_\Theta L_a(\vartheta, e, x) \, \xi_a(x, d\vartheta);$$

one obtains:

(4.8) Theorem

> *For a general Bayesian sequential decision problem let, for each $a \in \mathcal{A}, \xi_a$ be an a posteriori-distribution and $\varphi_a^\star$ be a terminal decision function such that*
>
> $$\int_\Theta \int_D L_a(\vartheta, e, x) \, \varphi_a^\star(x, de) \, \xi_a(x, d\vartheta) = \inf_{e \in D} \int_\Theta L_a(\vartheta, e, x) \, \xi_a(x, d\vartheta)$$
>
> *(and $\varphi_a^\star \equiv 1$ elsewhere). Then $\varphi^\star = (\varphi_a^\star)_{a \in A}$ is a Bayesian terminal decision procedure.*

Proof: Let $a \in \mathcal{A}$ and φ be an arbitrary terminal decision procedure; then

$$R(\xi, (a, \varphi^\star)) =$$
$$= \int_\Theta \int_\mathcal{X} \int_D \left(\int L_a(\vartheta, e, x) \varphi_a^\star(x, de) + C(\vartheta, a) \right) dP_\vartheta(x) \, d\xi(\vartheta)$$
$$\overset{(4.7)}{=} \int_\mathcal{X} \int_\Theta \int_D \left(\int L_a(\vartheta, e, x) \, \varphi_a^\star(x, de) + C(\vartheta, a) \right) \xi_a(x, d\vartheta) \, dP_\xi(x)$$
$$\overset{(assumption)}{=} \int_\mathcal{X} \inf_{e \in D} \int_\Theta \left(L_a(\vartheta, e, x) + C(\vartheta, a) \right) \xi_a(x, d\vartheta) \, dP_\xi(x)$$
$$\leq \int_\mathcal{X} \int_\Theta \int_D \left(\int L_a(\vartheta, e, x) \, \varphi_a(x, de) + C(\vartheta, a) \right) \xi_a(x, d\vartheta) \, dP_\xi(x)$$
$$= R(\xi, (a, \varphi));$$

(4.4) yields the assertion. $\qquad\square$

To illustrate these rather abstract assertions we again consider the simple examples (4.1) and (4.2):

(4.9) Example (Acceptance/rejection inspection "without replacement"; see (4.2))
In the situation of (4.2) let ξ be an a priori-distribution on $\mathcal{T} = \mathcal{P}(\Theta)$. Then

$$\xi_a(x,\{j\}) := \begin{cases} \dfrac{\xi(\{j\})P_j(\{y \in \mathcal{X} : y_i = x_i, 1 \leq i \leq g(a)\})}{P_\xi(\{y \in \mathcal{X} : y_i = x_i, 1 \leq i \leq g(a)\})} \\ \qquad\qquad \text{if the denominator is positive} \\ \xi(\{j\}) \qquad\qquad \text{elsewhere} \end{cases}$$

yields an a posteriori-distribution of ϑ under $\mathcal{B}_a, a \in \mathcal{A}$: Since the distributions are discrete the construction of ξ_a according to (4.6) reduces to solving, for $a \in \mathcal{A}$, the equation

$$P_j(\{y \in \mathcal{X} \ : \ y_i = x_i, \ 1 \leq i \leq g(a)\})\, \xi(\{j\}) =$$
$$= \xi_a(x,\{j\})\, P_\xi(\{y \in \mathcal{X} : y_i = x_i, 1 \leq i \leq g(a)\}).$$

Using for $x = (x_1, \ldots, x_m) \in \mathcal{X}$ and $0 \leq n \leq m$ the notation

$$S_n(x) = \sum_{i=1}^{n} x_i$$

one obtains for $0 \leq j \leq m$

$$P_j(\{y \in \mathcal{X} : y_i = x_i, 1 \leq i \leq g(a)\}) = \begin{cases} 0, \text{ if } S_{g(a)}(x) > j \text{ or } S_{g(a)}(x) + m - g(a) < j \\ \binom{m-g(a)}{j-S_{g(a)}(x)} \Big/ \binom{m}{j} \text{ elsewhere.} \end{cases}$$

This expression and therefore the a posteriori-distribution depend on a by $g(a)$ and on x by $S_{g(a)}(x)$ only. Hence there exists a representation

$$\xi_a(x,\cdot) = \hat{\xi}_{g(a)}(S_{g(a)}(x),\cdot)$$

where $\hat{\xi}_n(j,\cdot) : \mathcal{T} \to [0,1], 0 \leq n,j \leq m$.
The a posteriori-risks for the two terminal decisions "acceptance" and "rejection" are

$$\int_\Theta L_a(\vartheta, \text{acceptance}, x)\xi_a(x,d\vartheta) = \sum_{j=S_{g(a)}(x)}^{m-g(a)+S_{g(a)}(x)}(j - S_{g(a)}(x))\hat{\xi}_{g(a)}(S_{g(a)}(x),\{j\}) \text{ and}$$

$$\int_\Theta L_a(\vartheta, \text{rejection}, x)\xi_a(x,d\vartheta) = k_r(m - g(a)) \quad \text{resp.}$$

According to (4.8) one therefore obtains by

$$\varphi_a^*(x,\{\text{acceptance}\}) = \begin{cases} 1 \qquad\qquad\qquad\qquad\qquad\qquad\qquad\qquad\qquad\qquad\quad \leq \\ \text{if } \displaystyle\sum_{j=S_{g(a)}(x)}^{m-g(a)+S_{g(a)}(x)} (j - S_{g(a)}(x))\hat{\xi}_{g(a)}(S_{g(a)}(x),\{j\}) \qquad k_r(m-g(a)) \\ 0 \qquad\qquad\qquad\qquad\qquad\qquad\qquad\qquad\qquad\qquad\quad > \end{cases}$$

a Bayesian terminal decision procedure φ^*. To construct a Bayes-procedure it remains to solve the problem of optimal sampling for $(Z_a)_{a \in A}$ where

$$Z_a(x) := -\min\{k_r(m - S_{g(a)}(x)), \sum_{j=S_{g(a)}(x)}^{m-g(a)+S_{g(a)}(x)} (j - S_{g(a)}(x))\hat{\xi}_{g(a)}(S_{g(a)}(x), \{j\})\}$$
$$- c_0\, h(a) - cg(a).$$

According to (2.5) this can be done by backward induction. Due to the special structure of Z_a some further simplifications are possible: Using the notation

$$g_{\ell,i}(x):= -\min\{k_r(m - S_i(x)), \sum_{j=S_i(x)}^{m-i+S_i(x)} (j - S_i(x))\hat{\xi}_i(S_i(x), \{j\})\} - c_0\ell - c \cdot i$$

for $0 \le \ell k, \ell \le i \le m$, one obviously obtains

$$Z_a(x) = g_{h(a),g(a)}(x).$$

Also for the quantities U_a, defined according to (2.5), one may use versions which depend on a by $h(a)$ and $g(a)$ only: Defining for all i s.t. $k \le i \le m$

$$w_{k,i}(x) := g_{k,i}(x)$$

and

$$w_{\ell,i}(x) := \max_{i+1 \le t \le m} \{g_{\ell,i}(x), E_\xi(w_{\ell+1,t} \mid B_i)(x)\}$$

for $1 \le \ell < k, \ell \le i \le m$, and $\ell = i = 0$, it follows immediately from the definition of U_a that the $w_{h(a),g(a)}$ are versions of U_a. The quantities $g_{\ell,i}$ and $w_{\ell,i}$ resp. are identical with the terms used by Schüler [Schü] and by Abel [Ab 2]. Therefore the Bayesian sampling plan $\tau_{()}^*$ constructed according to (2.5) coincide with the inspection plan $\gamma^* = \{\gamma_j^*\}$ defined by Schüler (and Abel). $\qquad\square$

(4.10) **Example** (Acceptance/rejection sampling "with replacement"; see (4.1))

In the situation of (4.1) let the a priori-distribution be a beta-distribution with parameters $s, f > 0$, i.e. let

$$\xi(B) = \frac{1}{\beta(s, f)} \int_B \vartheta^{s-1}(1 - \vartheta)^{f-1}d\vartheta \quad \forall B \in I\!B^1_{|[0,1]}$$

(for this problem compare also Chapter 7 of Halds monograph [Had]). If for a sample sequence a, $S_{g(a)}(x)$ defective and therefore $F_{g(a)}(x) := g(a) - S_{g(a)}(x)$ effective items are found then the a posteriori-distribution is a beta-distribution with parameters $s + S_{g(a)}(x), f + F_{g(a)}(x)$ (see [Had], p. 130). Hence the a posteriori-risk of the terminal decision "acceptance" turns out to be

$$\int_\Theta L_a(\vartheta, \text{acceptance}, x), \xi_a(x, d\vartheta) =$$

$$= \frac{1}{\beta(s + S_{g(a)}(x), f + F_{g(a)}(x))} \cdot \int_0^1 K \cdot \vartheta \cdot \vartheta^{(s+S_{g(a)}(x)-1)}(1 - \vartheta)^{(f+F_{g(a)}(x)-1)}d\vartheta$$

$$= K \frac{\beta(s + S_{g(a)}(x) + 1, f + F_{g(a)}(x))}{\beta(s + S_{g(a)}(x), f + F_{g(a)}(x))}.$$

112

According to (4.8) one therefore obtains as a Bayesian terminal decision function

$$\varphi_a^\star(x,\{\text{acceptance}\}) = \begin{cases} 1 & \text{for } K\dfrac{\beta(s + S_{g(a)}(x) + 1, f + F_{g(a)}(x))}{\beta(s + S_{g(a)}(x), f + F_{g(a)}(x))} \leq K_r. \\ 0 & > \end{cases}$$

Hence the construction of a Bayes-procedure leads to a problem of optimal sequential sampling for $(Z_a)_{a \in \mathcal{A}}$ where

$$Z_a(x) := -\min\{K_r, K\frac{\beta(s + S_{g(a)}(x) + 1, f + F_{g(a)}(x))}{\beta(s + S_{g(a)}(x), f + F_{g(a)}(x))}\} - c_0\, h(a) - c\, g(a).$$

The numbers $(S_{g(a)}, F_{g(a)})_{a \in \tilde{\mathcal{A}}}$, occuring in this pay-off process, build a (homogeneous) Markov process with respect to $(\mathcal{B}_a)_{a \in \tilde{\mathcal{A}}}$ and P_ξ: Using the notation

$$Q_j((r,t), \{(r+n, t+j-n)\}) := \begin{cases} \dbinom{j}{n} \dfrac{\beta(s+r+n, f+t+j-n)}{\beta(s+r, f+t)} & \\ & \text{for } 0 \leq n \leq j \\ 0 & \text{elsewhere}, \end{cases}$$

$r, t \in \mathbb{N}_0$, $1 \leq j \leq m$, we obtain for $x = (x_1, \ldots, x_m) \in \mathcal{X}$ and $a, aj \in \tilde{\mathcal{A}}$

$$\begin{aligned}
P_\xi(&\{(S_{g(aj)}, F_{g(aj)}) = (r,t)\} \mid \mathcal{B}_a)(x) = \\
&= \frac{P_\xi(\{y \in \mathcal{X} : y_i = x_i, 1 \leq i \leq g(a), S_{g(aj)}(y) = r, F_{g(aj)}(y) = t\})}{P_\xi(\{y \in \mathcal{X} : y_i = x_i, 1 \leq i \leq g(a)\})} \\
&= Q_j((S_{g(a)}(x), F_{g(a)}(x)), \{(r,t)\}).
\end{aligned}$$

Therefore we are, for this problem of optimal sequential sampling, in the Markov case (see Ch. II; § 4). This allows a considerable simplification for the explicit solution, since one only has to determine the "continuation regions" B_j. For the special case given by the constants

$$\begin{aligned}
& m = 20, \quad c = 1, \quad K_r = 300, \quad K = 1800, \\
& s = 1, \quad f = 7, \text{ and} \\
& (i)\ c_0 = 4 \quad (ii)\ c_0 = 4.5 \quad (iii)\ c_0 = 5
\end{aligned}$$

the corresponding computations were carried out[35]; the resulting Bayes-optimal inspection plans are described by the figures 4.1 - 4.3.

---

[35] Using a program due to Lübbert (1988)

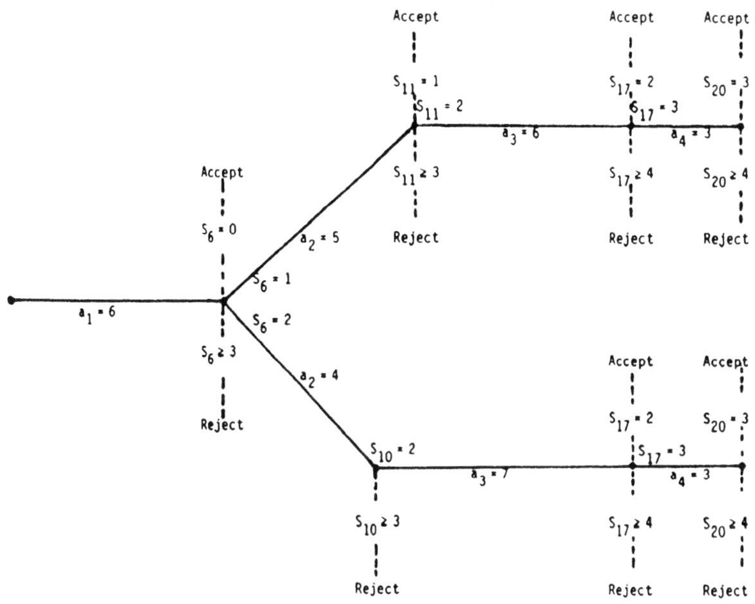

Figure 4.1
Bayes-optimal inspection plan for $c_0 = 4$

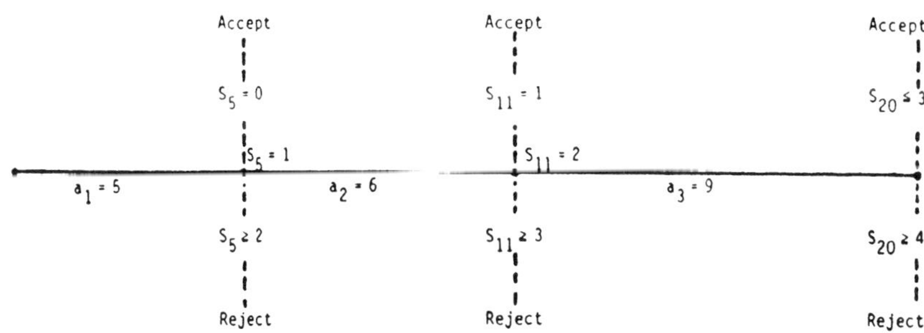

Figure 4.2
Bayes-optimal inspection plan for $c_0 = 4.5$

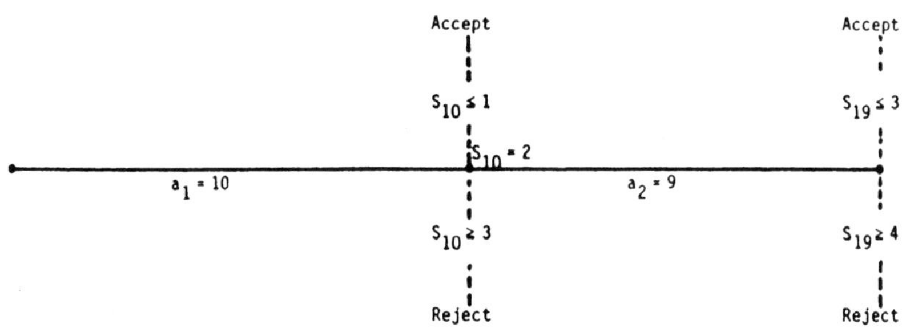

Figure 4.3

Bayes-optimal inspection plan for $c_0 = 5$

These results indicate a strong dependence of the optimal inspection plans on the fixed costs c_0. In particular, the optimal procedures become – as to be expected – "simpler" with increasing fixed costs c_0 – for $c_0 = 5$ already one obtains a simple two-stage test. This shows that the reserve/aversion of practical statisticians against purely sequential procedures (as mentioned in Ch. I, § 5) might be caused by good reasons, and that sequentially planned procedures are (here) the best possible ones.

§ 3 A posteriori-distributions

Theorem (4.8) and the remarks on p. 109 show that the a posteriori-distributions ξ_a play a key-role for the construction of Bayes-procedures. For the examples (4.9) and (4.10) these ξ_a were computed in an elementary way or were known resp.; in the sequel we give some general remarks on the construction of the ξ_a.

It is known (see e.g. [Ba], Theorem 56.5) that the *existence* of a posteriori-distributions can be proved if Θ is a complete separable metric space and $\mathcal{T}$ is the σ-algebra of Borel sets. Also from "classical" theory it is well-known (see e.g. [Wi 1], 2.7.3) that, under the assumption of $\mathcal{T} \otimes \mathcal{B}$-measurable densities, a posteriori-distributions can be constructed in an explicit way:

(4.11) Remark

Let the $P_\vartheta \mid \mathcal{B}_a, \vartheta \in \Theta$, be dominated by a σ-finite measure μ, let $f_{\vartheta,a}$ be densities such that

$$(\vartheta, x) \mapsto f_{\vartheta,a}(x)$$

is $T \otimes \mathcal{B}_a$-measurable, $a \in A$, and let ξ be an a priori-distribution. Using the notation

$$
\begin{aligned}
f_a(x) \quad &:= \int_\Theta f_{\vartheta,a}(x) \, d\xi(\vartheta) \\
h_a(\vartheta, x) \quad &:= \begin{cases} f_{\vartheta,a}(x)/f_a(x) & \text{if } f_a(x) \notin \{0, \infty\} \\ 1 & \text{elsewhere} \end{cases}
\end{aligned}
$$

one obtains by

$$
\xi_a(x, T) := \int_T h_a(\vartheta, x) \, d\xi(\vartheta), \quad T \in \mathcal{T}, \ x \in \mathcal{X},
$$

an a posteriori-distribution of ϑ under $\mathcal{B}_a$ (with the *a posteriori densitiy* $h_a(\cdot, x)$).

Using this "recipe" e.g. for the examples (4.9) and (4.10) immediately yields the a posteriori-distributions given there; for (4.10) one obtains (with the abbreviations $S(x) := S_{g(a)}(x)$, $F(x) := F_{g(a)}(x)$)

$$
\begin{aligned}
f_a(x) &= \int_0^1 \binom{g(a)}{S(x)} \vartheta^{S(x)}(1 - \vartheta)^{F(x)} \frac{\vartheta^{s-1}(1 - \vartheta)^{f-1}}{\beta(s, f)} \, d\vartheta \\
&= \binom{g(a)}{S(x)} \beta(s + S(x), f + F(x))/\beta(s, f), \\
h_a(x) &= \vartheta^{S(x)}(1 - \vartheta)^{F(x)} \beta(x, f)/\beta(s + S(x), f + F(x)),
\end{aligned}
$$

and

$$
\begin{aligned}
\xi_a(x, T) &= \int_T h_a(x) \, d\xi(\vartheta) \\
&= \frac{1}{\beta(s + S(x), f + F(x))} \int_T \vartheta^{s+S(x)-1}(1 - \vartheta)^{f+F(x)-1} \, d\vartheta,
\end{aligned}
$$

i.e. $\xi_a(x, \cdot)$ is a beta-distribution with parameters $s + S_{g(a)}(x)$, $f + F_{g(a)}(x)$. $\qquad \square$

For the example (4.10) follows, moreover, from the special structure (the single observations are independent) that we are in the Markov case, which considerably simplifies the construction of an explicit solution. Now we are going to show that this is caused by a more general argument:

(4.12) Let for the decision problem $\mathcal{P}$

$$
\begin{aligned}
\mathcal{A} &= A, \quad \mathcal{B}_{()} = \{\emptyset, \mathcal{X}\}, \\
\mathcal{B}_{(a_1,\ldots,a_n)} &= \sigma(Y_{(a_1)}, \ldots, Y_{(a_1,\ldots,a_n)}) \quad \forall \, (a_1, \ldots, a_n) \in A \backslash \{()\}
\end{aligned}
$$

where $(Y_a)_{a \in A \backslash \{()\}}$ is a family of random variables

$$
Y_{(a_1,\ldots,a_n)} : (\mathcal{X}, \mathcal{B}) \to (\mathcal{Y}_{a_n}, \mathcal{C}_{a_n})
$$

such that

(i) for each $(a_1, a_2, \ldots) \in A^*$ and $\vartheta \in \Theta$ $(Y_{(a_1,\ldots,a_n)})_{n \in \mathbb{N}}$ is a stochastically independent (with respect to P_ϑ) sequence of random variables

(ii) $P_\vartheta^{Y_a} = P_\vartheta^{Y_b} \quad \forall \, a = (a_1, \ldots, a_n), \ b = (b_1, \ldots, b_r) \in A \text{ s.t. } a_n = b_r$

116

(iii) $(P_\vartheta^{Y_{(j)}})_{\vartheta \in \Theta}$ is dominated by a σ-finite measure μ_j on $(\mathcal{Y}_j, \mathcal{C}_j)$ and for the densities $f_{\vartheta,j}$ of $P_\vartheta^{Y_{(j)}}$

$$(\vartheta, y) \mapsto f_{\vartheta,j}(y)$$

is $(\mathcal{T} \otimes \mathcal{C}_j, I\!\!B)$-measurable. $\qquad\qquad\square$

This situation is, in particular, given for sequential sampling plans with *independent identically distributed* single observations. Then the σ-algebra $\mathcal{B}_a$ is generated by iid random variables $X_1, \ldots, X_{g(a)}$ where $X_n : (\mathcal{X}, \mathcal{B}) \to (\mathcal{Y}, \mathcal{C}), n \in I\!\!N$. Defining $(\mathcal{Y}_j, \mathcal{C}_j) = (\mathcal{Y}, \mathcal{C})^j$ and

$$Y_{(a_1,\ldots,a_n)} := (X_{g((a_1,\ldots,a_{n-1}))+1}, \ldots, X_{g((a_1,\ldots,a_{n-1}))+a_n})$$

(i.e. as n-th "segment") a process yields $(Y_a)_{a \in A \setminus \{()\}}$ that fulfils the properties formulated in (4.12).

In the situation (4.12), a posteriori-distributions can explicitly be constructed by using (4.11). For the subsequent proof, that these a posteriori distributions form a Markov process, it is convenient to use a recursive representation:

(4.13) Lemma
   In the situation (4.12) let ξ be an a priori-distribution on Θ and let for $a \in A$

$$\xi_a : \mathcal{X} \times \mathcal{T} \to [0,1]$$

be defined by $\xi_{()}(x, T) := \xi(T)$ and

$$\xi_a(x, T) := \begin{cases} \dfrac{\displaystyle\int_T \prod_{i=1}^n f_{\vartheta,a_i}(Y_{(a_1,\ldots,a_i)}(x))d\xi(\vartheta)}{\displaystyle\int_\Theta \prod_{i=1}^n f_{\vartheta,a_i}(Y_{(a_1,\ldots,a_i)}(x))d\xi(\vartheta)} & \text{if the denominator } \notin \{0, \infty\} \\[4ex] \xi(T) & \text{elsewhere.} \end{cases}$$

Then ξ_a is an a posteriori-distribution of ϑ under $\mathcal{B}_a$ for ξ, $a \in A$, and it follows P_ξ-a.s. for $a \in A, j \in M$ s.t. $aj \in A$

$$\xi_{aj}(x, T) = \begin{cases} \dfrac{\displaystyle\int_T f_{\vartheta,j}(Y_{aj}(x))\xi_a(x, d\vartheta)}{\displaystyle\int_\Theta f_{\vartheta,j}(Y_{aj}(x))\xi_a(x, d\vartheta)} & \text{if the denominator } \notin \{0, \infty\} \\[4ex] \xi_a(x, T) & \text{elsewhere.} \end{cases}$$

Proof: The first assertion is a consequence of (4.11) and the assumption of (4.12), the second one follows from the definition of ξ_a. $\qquad\qquad\square$

The "independence structure" of (4.12) yields for the a posteriori-distributions:

(4.14) Theorem ([Lü], (1.4.2))
   In the situation (4.12) let ξ be an a priori-distribution on Θ, and let $\xi_a, a \in A$, be the a posteriori-distributions for ξ defined in (4.13). Let W be the family of all

probability measures on $(\Theta, \mathcal{T})$ and $\mathcal{W}$ the smallest σ-algebra on W such that all mappings $\rho \mapsto \rho(T)$, $T \in \mathcal{T}$, are measurable, and let P_ρ, for $\rho \in W$, be the probability measure on $(\mathcal{X}, \mathcal{B})$ defined by

$$P_\rho(B) := \int_\Theta P_\vartheta(B)\, d\rho(\vartheta)$$

(see p. 109). Then it follows for $\zeta_a : \mathcal{X} \to W$ defined by

$$\zeta_a(x) := \xi_a(x, \cdot),\ a \in A :$$

$(\zeta_a)_{a\in A}$ *is a stationary Markov-process with respect to* $(\mathcal{B}_a)_{a\in A}$ *and* P_ξ *with initial state ξ and transition probabilities (stochastic stic kernels) $Q_j, j \in M$, which may be chosen independently of ξ.*

Proof ([Lü], p. 31 ff): (i) ζ_a is $(\mathcal{B}_a, \mathcal{W})$-measurable.
A mapping u from a measurable space $(\mathcal{Y}, \mathcal{C})$ to $(W, \mathcal{W})$ is $(\mathcal{C}, \mathcal{W})$-measurable iff

$$K : \mathcal{Y} \times \mathcal{T} \to [0, 1],\ K(y, T) := u(y)(T),$$

defines a stochastic kernel from $(\mathcal{Y}, \mathcal{C})$ to $(\Theta, \mathcal{T})$ (see [Ba], Theorem 7.4). Since ξ_a is a stochastic kernel from $(\mathcal{X}, \mathcal{B}_a)$ to $(\Theta, \mathcal{T})$, assertion (i) follows.

(ii) For $j \in M$ let $h_j : W \times \mathcal{Y}_j \to W$ be defined by

$$h_j(\rho, y)(T) := \begin{cases} \dfrac{\displaystyle\int_T f_{\vartheta,j}(y)d\rho(\vartheta)}{\displaystyle\int_\Theta f_{\vartheta,j}(y)d\rho(\vartheta)} & \text{if the denominator } \notin \{0, \infty\} \\[4ex] \rho(T) & \text{elsewhere.} \end{cases}$$

Then h_j is $(\mathcal{W}\otimes\mathcal{C}_j, \mathcal{W})$- measurable. According to the definition of $\mathcal{W}$ it is (analogously to the proof of (i)) sufficient to show that the mapping

$$(\rho, y) \mapsto h_j(\rho, y)(T)$$

is $(\mathcal{W}\otimes\mathcal{C}_j, I\!B)$-measurable $\forall T \in \mathcal{T}$. For this purpose we mention that the mapping

$$(\rho, y) \mapsto \int_\Theta 1_{T\times B}(\vartheta, y)d\rho(\vartheta) = \rho(T)\, 1_B(y)$$

is $(\mathcal{W}\otimes\mathcal{C}_j, I\!B)$-measurable $\forall T \in \mathcal{T}, B \in \mathcal{C}_j$. Hence it follows by algebraic induction that the mapping

$$(\rho, y) \mapsto \int_\Theta g(\vartheta, y)d\rho(\vartheta)$$

is $(\mathcal{W}\otimes\mathcal{C}_j, I\!B)$-measurable for each $(\mathcal{T}\otimes\mathcal{C}_j, I\!B)$-measurable mapping $g : \Theta \times Y_j \to [0, \infty)$, in particular for $f_{\vartheta,j}(y)\, 1_T(\vartheta)$.

(iii) For $j \in M$ let $Q_j : W \times \mathcal{W} \to [0, 1]$ be defined by

$$Q_j(\rho, V) := P_\rho(\{x \in \mathcal{X} : h_j(\rho, Y_{(j)}(x)) \in V\}).$$

118

Then Q_j is a transition probability from $(W, \mathcal{W})$ to $(W, \mathcal{W})$.

For each $\rho \in W$ the mapping $Q_j(\rho, \cdot)$ obviously yields a probability measure on $(W, \mathcal{W})$. Therefore it remains to show that the mapping

$$\rho \mapsto Q_j(\rho, V)$$

is measurable for each $V \in \mathcal{W}$. For this aim we first prove that the mapping

$$(\vartheta, \rho) \mapsto P_\vartheta(\{x \in \mathcal{X} : (x, \rho) \in C\})$$

is $(\mathcal{T} \otimes \mathcal{W}, \mathbb{B})$-measurable for each $C \in \mathcal{B} \otimes \mathcal{W}$: For all sets C of the form $C = F \times V$ where $F \in \mathcal{B}, V \in \mathcal{W}$, measurability holds (because of the measurability of $\vartheta \mapsto P_\vartheta(F)$). Since these sets form a $\cap$-stable system which generates $\mathcal{B} \otimes \mathcal{W}$, the desired measurability follows. In particular for

$$C := \{(x, \rho) : h_j(\rho, Y_{(j)}(x)) \in V\}$$

one therefore obtains from the measurability of h_j that

$$(\vartheta, \rho) \mapsto P_\vartheta(\{x \in \mathcal{X} : h_j(\rho, Y_{(j)}(x)) \in V\})$$

is $(\mathcal{T} \otimes \mathcal{W}, \mathbb{B})$-measurable. Since on the other hand

$$(\rho, T) \mapsto \rho(T)$$

yields a stochastic kernel from $(W, \mathcal{W})$ to $(\Theta, \mathcal{T})$ it follows, using Fubini's theorem, that

$$\rho \mapsto \int_\Theta P_\vartheta(\{x \in \mathcal{X} : h_j(\rho, Y_{(j)}(x)) \in V\}) d\rho(\vartheta) = Q_j(\rho, V)$$

is measurable for all $V \in \mathcal{W}$.

(iv) $P_\xi(\zeta_{aj} \in V \mid \mathcal{B}_a) = Q_j(\zeta_a, V)\ P_\xi - a.s.$ for all $a \in A, j \in M$ and $V \in \mathcal{W}$. From (i) and (iii) the $(\mathcal{B}_a, \mathbb{B})$-measurability of $Q_j(\zeta_a, V)$ follows. For each $F \in \mathcal{B}_a$ one obtains

$$\int_F 1_V(\zeta_{aj}) dP_\xi = \int_F 1_V(h_j(\zeta_a, Y_{aj})) dP_\xi$$

$$\text{(according to the definition of } \zeta_a \text{ (see (4.13))}$$

$$= \int_\Theta \int_F 1_V(h_j(\zeta_a, Y_{aj}))\, dP_\vartheta\, d\xi(\vartheta)$$

$$\text{(according to the definition of } P_\xi)$$

$$= \int_\Theta \int_F E_\vartheta(1_V(h_j(\zeta_a, Y_{aj})) \mid \mathcal{B}_a) dP_\vartheta\, d\xi(\vartheta)$$

$$\text{(since } F \in \mathcal{B}_a)$$

$$= \int_\Theta \int_F P_\vartheta(\{\tilde{x} \in \mathcal{X} : h_j(\zeta_a(x), Y_{(j)}(\tilde{x})) \in V\}) dP_\vartheta(x) d\xi(\vartheta)$$

$$\text{(since } Y_{aj} \text{ is independent of } \mathcal{B}_a \text{ and } P_\vartheta^{Y_{aj}} = P_\vartheta^{Y_{(j)}})$$

$$= \int_F \int_\Theta P_\vartheta(\{\tilde{x} \in \mathcal{X} : h_j(\zeta_a(x), Y_{(j)}(\tilde{x})) \in V\}) \xi_a(x, d\vartheta)\, dP_\xi(x)$$

$$\text{(according to (4.7))}$$

$$= \int_F Q_j(\zeta_a, V)\, dP_\xi$$

and therefore the assertion.

Moreover, one obtains

$$
\begin{aligned}
P_\xi(\zeta_{aj} \in V \mid \mathcal{B}_a) &= Q_j(\cdot, V) \circ \zeta_a = E_\xi(Q_j(\cdot, V) \circ \zeta_a \mid \zeta_a) \\
&= E_\xi(P_\xi(\zeta_{aj} \in V \mid \mathcal{B}_a) \mid \zeta_a) = P_\xi(\zeta_{aj} \in V \mid \zeta_a)\, P_\xi - a.s.,
\end{aligned}
$$

i.e. the Markov property. $\qquad\square$

(4.15) Remarks

a) In many problems – particularly for conjugate a priori-distributions (see [DeG], Ch. 9) – all a posteriori-distributions which can occur for a given a priori-distribution belong to a (small) subset $\tilde{W}$ of W. The assertions of theorem (4.14) remain valid if $(\zeta_a)_{a \in A}$ is considered as stochastic process with values in $(\tilde{W}, \mathcal{W}_{\tilde{W}})$.

b) Theorem (4.14) was only concerned with versions of a posteriori-distributions as defined in (4.13). But if the σ-algebra $\mathcal{T}$ on Θ is countably generated, different versions of ξ_a coincide P_ξ-a.s.; hence arbitrary versions of the a posteriori-distributions form a stationary Markov-process with transition probabilities Q_j, $j \in M$, defined in (4.14).

c) Because of $\mathcal{B}_{()} = \{\emptyset, \mathcal{X}\}$ the distribution $Q_j(\xi, \cdot)$ is uniquely determined.

§ 4 Bayes-optimal sampling plans; Markov case

In § 2 we have shown that the construction of Bayes-optimal sequentially planned decision procedures may be split up into two nearly independent steps, namely
– to determine a Bayes-optimal terminal decision procedure $\varphi^\star = (\varphi_a^\star)_{a \in A}$, and
– to solve the problem of optimal sampling for

$$
Z_a = \begin{cases}
- \int_\Theta (\int_D L_a(\vartheta, e, x)\varphi_a^\star(x, de) + C(\vartheta, a))\xi_a(x, d\vartheta) & \text{for } a \in \mathcal{A} \\
-\infty & \text{elsewhere.}
\end{cases}
$$

The first step was extensively investigated in the "classical" theory (see e.g. [Be], Ch. 4.4; [B/P]); in the sequel we assume that a Bayes-optimal terminal decision procedure is given. Then we have to determine a Bayes-optimal sampling plan.

For purely sequential decision problems – i.e. for the special case $M = \{1\}$ – under the assumption that
– the conditions of (4.12) are fulfilled,
– the loss function $(L_a)_{a \in A}$ is independent of a and x, and
– the sampling costs grow linearly, i.e. $C(\vartheta, n) = c \cdot n$,

one can describe the solution of the problem of optimal sampling in the following way (see [DeG], Ch. 12): Let ξ_n be the a posteriori-distribution at the "time point" $n \in \mathbb{N}$. Denoting by

$$
R_1(\xi) := \inf\{R(\xi, (\tau, \varphi)) : \tau \succ ()\}
$$

the minimal Bayes-risk of all procedures requiring at least one observation, and by

$$
\chi(\xi) := \inf\{R(\xi, (\tau, \varphi) : \tau \equiv ()\}
$$

120

the minimal Bayes-risk of all procedures making a terminal decision without any observation, then for

$R_1(\xi_n) > \chi(\xi_n)$ a Bayes-optimal terminal decision has to be made

$R_1(\xi_n) < \chi(\xi_n)$ a further observation is required

(for $R_1(\xi_n) = \chi(\xi_n)$ both possibilities are equally good).

Despite the fact that the situation is much more complex for general sequential decision problems, we shall prove, under corresponding assumptions on the loss function and the sampling costs, similar results.

(4.16) Let the situation of (4.12) be given, and assume that

(i) $L_a(\vartheta, e, x) = L(\vartheta, e) \; \forall a \in A, \; \vartheta \in \Theta, e \in D, x \in \mathcal{X}$, i.e. the loss function depends only on the underlying distribution and the decision but neither on the sample sequence nor on the special sample,

(ii) ξ is an a priori-distribution on $(\Theta, \mathcal{T})$,

(iii) there exists a subset $\tilde{W}$ of W such that

    a) $\xi \in \tilde{W}$ and $\xi_a^{(\rho)}(x, \cdot) \in \tilde{W} \; \forall a \in A, x \in \mathcal{X}, \rho \in \tilde{W}$ where $\xi_a^{(\rho)}, a \in A$, denotes the versions of the a posteriori-distributions of ϑ under $\mathcal{B}_a$ given by (4.13),

    b) for each $\rho \in \tilde{W}$ there exists a Bayes-optimal terminal decision procedure $\varphi^{(\rho)}$,

    c) the mapping $\psi : (\tilde{W}, \mathcal{W}_{\tilde{W}}) \to (\overline{\mathbb{R}}, \overline{\mathbb{B}})$ defined by

$$\psi(\rho) := \inf_{e \in D} \int_\Theta L(\vartheta, e) \; \rho(d\vartheta)$$

    is measurable,

    d) for

$$Z_a^{(\rho)}(x) := -\psi(\xi_a^{(\rho)}(x, \cdot)) - \int_\Theta C(\vartheta, a)\xi_a^{(\rho)}(x, d\vartheta)$$

    (see p. 109) holds

$$E_\rho \, Z_a^{(\rho)} > -\infty \quad \forall a \in A, \rho \in \tilde{W}.$$

The assumptions (4.16) are straightforward generalizations of the corresponding assumptions for the purely sequential case to our more complicated situation – (iii) a)-d) make "technically" precise what is meant by the assumed existence of a solution for the first of the two steps for constructing a Bayes-solution.

For the finite horizon case one obtains

(4.17) Theorem ([Lü], (1.4.10))

Let the assumptions (4.16) be fulfilled; for $m \in \mathrm{IN}, j \in M$ s.t. $j \leq m$ and $\rho \in \tilde{W}$ define

$$\begin{aligned}
B^m(\rho) &:= \inf\{R(\rho, (\tau, \varphi^{(\rho)})) : g(\tau) \leq m\}, \\
B_j^m(\rho) &:= \inf\{R(\rho, (\tau, \varphi^{(\rho)})) : g(\tau) \leq m, \tau \succeq (j)\}.
\end{aligned}$$

Then it follows for the problem of optimal sequential sampling with pay-offs $Z_a^{(\rho)}, a \in A^m$

$$U_a^{\rho,m}(x) = -B^{m-g(a)}(\zeta_a^{(\rho)}(x)) - \int_\Theta C(\vartheta, a)\xi_a^{(\rho)}(x, d\vartheta),$$

$$E_\rho(U_{aj}^{\rho,m} \mid \mathcal{B}_a)(x) = -B_j^{m-g(a)}(\zeta_a^{(\rho)}(x)) - \int_\Theta C(\vartheta, a)\xi_a^{(\rho)}(x, d\vartheta)$$

(comp. (2.24)). The B_j^m fulfill the Bellman-equations

$$\begin{aligned}
B^m(\rho) &= \min\{\psi(\rho), B_j^m(\rho) : j \in M, j \leq m\}, \\
B_j^m(\rho) &= \int_{\tilde{W}} B^{m-j}(\pi)Q_j(\rho, d\pi) + \int_\Theta c(\vartheta, j)d\rho(\vartheta),
\end{aligned}$$

where Q_j are the transition probabilities of (4.14) and $B^o(\rho) := \psi(\rho)$.

Proof: According to (4.14)/(4.16) the assumptions of (2.25) are fulfilled with

$$(\mathcal{Y}, \mathcal{C}) = (\tilde{W}, \mathcal{W}_{\tilde{W}}), \quad Y_a^{(\rho)} := \zeta_a^{(\rho)},$$

and

$$\eta_a : (\mathcal{Y}, \mathcal{C}) \to (\overline{\mathbb{R}}, \overline{\mathbb{B}}) \text{ defined by } \eta_a(\rho) := -\psi(\rho) - \int_\Theta C(\vartheta, a)d\rho(\vartheta).$$

Hence one obtains from (2.25)

$$E_\rho(U_{aj}^{\rho,m} \mid \mathcal{B}_a) = v_{a,j,m-g(a)}(\zeta_a^{(\rho)}) \quad P_\rho - a.s..$$

For each $\tau \in T_{(j)}^{m-g(a)}$ follows

$$\begin{aligned}
E_\rho\eta_{a\tau}(\zeta_\tau^{(\rho)}) &= -E_\rho(\psi(\zeta_\tau^{(\rho)}) + \int_\Theta C(\vartheta, a\tau)\xi_\tau^{(\rho)}(\cdot, d\vartheta)) \\
&= -E_\rho(\psi(\zeta_\tau^{(\rho)}) + \int_\Theta C(\vartheta, \tau)\xi_\tau^{(\rho)}(\cdot, d\vartheta)) \\
&\quad - E_\rho(\int_\Theta C(\vartheta, a)\xi_\tau^{(\rho)}(\cdot, d\vartheta)) \\
&\quad \text{(according to the definition of } C(\vartheta, \cdot)) \\
&\overset{(4.7)}{=} -E_\rho\eta_\tau(\zeta_\tau^{(\rho)}) - \int_\Theta C(\vartheta, a)d\rho(\vartheta).
\end{aligned}$$

Taking the supremum with respect to τ yields

$$v_{a,j,m-g(a)}(\rho) = -B_j^{m-g(a)}(\rho) - \int_\Theta C(\vartheta, a)d\rho(\vartheta)$$

and therefore altogether

$$E_\rho(U_{aj}^{\rho,m} \mid \mathcal{B}_a) = -B_j^{m-g(a)}(\zeta_a^{(\rho)}) - \int_\Theta C(\vartheta, a)\xi_a^{(\rho)}(\cdot, d\vartheta) \quad P_\rho - a.s..$$

Obviously

$$B^m(\rho) = \min\{\psi(\rho), B_j^m(\rho) : j \in M, j \leq m\};$$

122

hence

$$
\begin{aligned}
U_a^{\rho,m} &= \max\{Z_a^{(\rho)}, E_\rho(U_{aj}^{\rho,m} \mid \mathcal{B}_a) : j \in M, j \le m - g(a)\} \\
&= -\min\{\psi(\zeta_a^{(\rho)}), B_j^{m-g(a)}(\zeta_a^{(\rho)}) : j \in M, j \le m - g(a)\} \\
&\quad - \int_\Theta C(\vartheta, a)\xi_a^{(\rho)}(\cdot, d\vartheta) \\
&= -B^{m-g(a)}(\zeta_a^{(\rho)}) - \int_\Theta C(\vartheta, a)\xi_a^{(\rho)}(\cdot, d\vartheta).
\end{aligned}
$$

Finally we obtain

$$
\begin{aligned}
B_j^m(\rho) &= -E_\rho(U_{(j)}^{\rho,m} \mid \mathcal{B}_{()}) \quad (\text{special case } a = ()) \\
&= E_\rho\Big(B^{m-j}(\zeta_{(j)}^{(\rho)}) + \int_\Theta c(\vartheta, j)\xi_{(j)}^{(\rho)}(\cdot, d\vartheta) \mid \mathcal{B}_{()}\Big) \\
&\quad (\text{according to the previous result for } a = (j)) \\
&= \int_{\tilde{W}}\Big(B^{m-j}(\pi) + \int_\Theta c(\vartheta, j)d\pi(\vartheta)\Big)\, Q_j(\rho, d\pi) \\
&\quad (\text{since } Q_j \text{ is a transition probability for } j \in M) \\
&= \int_{\tilde{W}} B^{m-j}(\pi)\, Q_j(\rho, d\pi) + \int_{\tilde{W}}\int_\Theta c(\vartheta, j)\, d\pi(\vartheta)P_\rho^{\xi_{(j)}^{(\rho)}}(d\pi) \\
&\quad (\text{according to the definition of } Q_j) \\
&= \int_{\tilde{W}} B^{m-j}(\pi)\, Q_j(\rho, d\pi) + \int_X\int_\Theta c(\vartheta, j)\xi_{(j)}^{(\rho)}(x, d\vartheta)dP_\rho(x) \\
&\stackrel{(4.7)}{=} \int_{\tilde{W}} B^{m-j}(\pi)\, Q_j(\rho, d\pi) + \int_\Theta\int_X c(\vartheta, j)\, dP_\vartheta(x)\, d\rho(\vartheta) \\
&= \int_{\tilde{W}} B^{m-j}(\pi)\, Q_j(\rho, d\pi) + \int_\Theta c(\vartheta, j)\, d\rho(\vartheta). \qquad \square
\end{aligned}
$$

Theorem (4.17) together with (2.5) yields the following result: At "decision time" $a \in A$ let ξ_a (according to (4.13)) be the a posteriori-distribution and let

$$
R(\xi_a) := \min\{B_j^{m-g(a)}(\zeta_a) : j \in M, j \le g(a) - m\}.
$$

Then $(\tau^\star, \varphi^{(\xi)})$, where the sampling plan $\tau^\star$ is defined by taking

$$
\text{a terminal decision according to } \varphi_a^{(\xi)} \text{ if } R(\xi_a) \ge \psi(\zeta_a)
$$

and

a further sub-sample of a size $j^\star$ for which $B_{j^\star}^{m-g(a)}(\zeta_a) = R(\xi_a)$,
if $R(\xi_a) < \psi(\zeta_a)$,

is a Bayes-procedure with respect to ξ.

The Bayes-procedures, given for example (4.10) ("Acceptance/rejection sampling with replacement"), have been constructed in just that way – $\zeta_a(x)$ is the beta-distribution with parameters $s + S_{g(a)}(x)$ and $f + F_{g(a)}(x)$, $\varphi_a^{(\xi)}$ corresponds to $\varphi_a^\star$ given there, and the $B_j^{m-g(a)}$ are computed by backward induction.

Our next goal is to show that for an infinite horizon

$$
\lim_{m \to \infty} B^n(\xi) = B(\xi) := \inf_{(\tau,\varphi)} R(\xi, (\tau, \varphi)),
$$

i.e. that the minimal Bayes-risk for an infinite horizon may be approximated by the risks for a finite horizon. For this aim we give a condition which guarantees that, for a problem of optimal stopping,

$$\lim_{m \to \infty} U_a^m = U_a \quad P-a.s.$$

for the values U_a^m corresponding to the sets $\mathcal{A}^m := \{a \in \mathcal{A} : g(m) \le m\}$. Since the U_a^m are non-decreasing in m and bounded from above by U_a, one obtains that

$$U_a' := \lim_{m \to \infty} U_a^m$$

exists for each $a \in \tilde{A}$ and that $U_a' \le U_a$.

(4.18) Lemma

Define for $a = (a_1, a_2, \ldots) \in A \cup A^*$ and $m \in \mathrm{IN}$

$$a^{(m)} := \left\{ \begin{array}{ll} a & \text{if } g(a) \le m \\ (a_1, \ldots, a_j) & \text{if } j = \max\{k : g(a_1, \ldots, a_k) \le m\}, \end{array} \right.$$

and assume that

$$\lim_{m \to \infty} \int_{\{g(\tau) > m\}} U_{\tau^{(m)}}' \, dP = 0$$

for each $\tau \in T_{()}$. Then $U_a' = U_a \quad \forall a \in \tilde{A}$.

Proof: Since according to (2.5)

$$U_a^m = \max\{Z_a, E(U_{aj}^m \mid \mathcal{B}_a) : j \in \tilde{M}_a, g(aj) \le m\}, \text{ if } g(a) \le m,$$

and U_a^m converges (for $m \to \infty$) monotonically to U_a', the monotone convergence theorem for conditional expectations yields

$$U_a' = \sup\{Z_a, E(U_{aj}' \mid \mathcal{B}_a) : j \in \tilde{M}_a\}.$$

This means that $(U_a')_{a \in \mathcal{A}}$ is a supermartingale. For $\tau \in T_a$ one obtains from

$$E(\sup_{b \in \tilde{A}} Z_b \mid \mathcal{B}_a) \ge U_a' \ge Z_a \text{ and } E(\sup_{b \in \tilde{A}} Z_b) < \infty$$

that $E \mid U_\tau' \mid < \infty$; moreover, by assumption

$$\liminf_{m \to \infty} \int_{\{g(\tau) > m\}} U_{\tau^{(m)}}' \, dP = 0.$$

Hence the optional sampling theorem (see e.g. [Wa/Wi]) yields

$$U_a' \ge E(U_\tau' \mid \mathcal{B}_a) \ge E(Z_\tau \mid \mathcal{B}_a),$$

i.e. the needed inequality $U_a' \ge U_a$. $\qquad\square$

This general lemma will now be used for a proposition on Bayes-optimal sampling plans:

124

(4.19) Lemma

Let the situation (4.12) be given and let ξ be an a priori- distribution on (Θ, T); assume that the loss functions L_a fulfill

$$L_a(\vartheta, e, x) = L(\vartheta, e) \; \forall a \in A, \vartheta \in \Theta, \; e \in D, \; x \in \mathcal{X}$$

(i.e. (4.16)i)), and let $\varphi_a^\star$ be a Bayesian terminal decision procedure. If for

$$Z_a(x) := -\int_\Theta \left(\int_D L(\vartheta, e) \varphi_a^\star(x, de) + C(\vartheta, a) \right) \xi_a(x, d\vartheta)$$

$$\int_{\mathcal{X}} Z_a \, dP_\xi > -\infty \quad \forall a \in A \text{ then } \lim_{m \to \infty} U_a^m = U_a.$$

Proof: Because of $0 \geq U_a' \geq Z_a$ it is sufficient to show that

$$\lim_{m \to \infty} \int_{\{g(\tau) > m\}} Z_{\tau(m)} dP_\xi = 0$$

for each $\tau \in T_()$; then (4.18) yields the assertion.

For this aim we consider the two summands of Z_a separately: For $m \in \mathrm{IN}$

$$\begin{aligned}
0 \;\leq\;& \int_{\{g(\tau) > m\}} \int_\Theta C(\vartheta, \tau^{(m)}) \xi_{\tau(m)}(x, d\vartheta) \, dP_\xi(x) \\
=\;& \int_\Theta \int_{\{g(\tau) > m\}} C(\vartheta, \tau^{(m)}) \, dP_\vartheta(x) \, d\xi(\vartheta) \text{ according to (4.7)} \\
\leq\;& \int_\Theta \int_{\{g(\tau) > m\}} C(\vartheta, \tau) \, dP_\vartheta(x) \, d\xi(\vartheta) \\
& \text{since } \tau \succeq \tau^{(m)} \text{ and } C \text{ is monotonic} \\
=\;& \int_{\{g(\tau) > m\}} \int_\Theta C(\vartheta, \tau) \, \xi_\tau(x, d\vartheta) \, dP_\xi(x) \text{ according to (4.7)} \\
\leq\;& \int_{\{g(\tau) > m\}} -Z_\tau(x) \, dP_\xi(x).
\end{aligned}$$

Since $E_\xi \mid Z_\tau \mid \; < \infty$ for $\tau \in T_()$ and

$$P_\xi(\{g(\tau) > m\}) \xrightarrow{m \to \infty} 0$$

the last term converges to 0 for $m \to \infty$. Analogously one obtains for the second summand

$$\begin{aligned}
0 \;\leq\;& \int_{\{g(\tau) > m\}} \int_\Theta \int_D L(\vartheta, e) \, \varphi_{\tau(m)}^\star(x, de) \, \xi_{\tau(m)}(x, d\vartheta) \, dP_\xi(x) \\
\leq\;& \int_{\{g(\tau) > m\}} \int_\Theta \int_D L(\vartheta, e) \, \varphi_()^\star(x, de) \, \xi_{\tau(m)}(x, d\vartheta) \, dP_\xi(x) \\
& \text{since the Bayesian terminal procedure } \varphi_{\tau(m)}^\star \text{ is replaced by } \varphi_()^\star \\
=\;& \int_\Theta \int_{\{g(\tau) > m\}} \int_D L(\vartheta, e) \, \varphi_()^\star(x, de) \, dP_\vartheta(x) \, d\xi(\vartheta) \\
& \text{according to (4.7)} \\
=\;& \int_{\{g(\tau) > m\}} \int_\Theta \int_D L(\vartheta, e) \, \varphi_()^\star(x, de) \, \xi_\tau(x, d\vartheta) \, dP_\xi(x) \\
& \text{again according to (4.7)}.
\end{aligned}$$

Since

$$\int_{X} \int_{\Theta} \int_{D} L(\vartheta, e)\, \varphi_{()}^{*}(x, de)\, \xi_{\tau}(x, d\vartheta)\, dP_{\xi}(x) = -\int_{X} Z_{()}(x)\, dP_{\xi}(x) < \infty$$

and $P_{\xi}(\{g(\tau) > m\}) \xrightarrow{m \to \infty} 0$ the last term converges (for $m \to \infty$) to 0, too. $\quad\square$

This yields for the Bayes-risks at an infinite horizon:

(4.20) Theorem

Let the assumptions (4.16) be fulfilled; for $\rho \in \tilde{W}$ and $j \in M$ define

$$\begin{aligned}
B(\rho) &:= \inf\{R(\rho, (\tau, \varphi^{(\rho)}))\}, \\
B_j(\rho) &:= \inf\{R(\rho, (\tau, \varphi^{(\rho)})) : \tau \succeq (j)\}.
\end{aligned}$$

Then

$$B(\rho) = \lim_{m \to \infty} B^m(\rho), \ B_j(\rho) = \lim_{m \to \infty} B_j^m(\rho).$$

The B_j fulfill the Bellman-equations

$$\begin{aligned}
B(\rho) &= \inf\{\psi(\rho),\ B_j(\rho) : j \in M\}, \\
B_j(\rho) &= \int_{\tilde{W}} B(\pi)\, Q_j(\rho, d\pi) + \int_{\Theta} c(\vartheta, j)\, d\rho(\vartheta), \rho \in \tilde{W}, j \in M,
\end{aligned}$$

where Q_j denotes the transition probabilities of (4.14). For the problem of optimal sequential sampling with the pay-off process $(Z_a^{(\rho)})_{a \in A}$

$$U_a^{(\rho)}(x) = -B(\zeta_a^{(\rho)}(x)) - \int_{\Theta} C(\vartheta, a)\, \xi_a^{(\rho)}(x, d\vartheta).$$

$$E_\rho(U_{aj}^{(\rho)} \mid \mathcal{B}_a)(x) = -B_j(\zeta_a^{(\rho)}(x)) - \int_{\Theta} C(\vartheta, a) \xi_a^{(\rho)}(x, d\vartheta) \quad P_\rho - a.s.$$

(see (2.24)).

Proof: According to (4.19) one obtains for each $a \in A$

$$\lim_{m \to \infty} U_a^{\rho, m} = U_a^{(\rho)};$$

hence

$$\lim_{m \to \infty} E_\rho(U_{aj}^{\rho, m} \mid \mathcal{B}_a) = E_\rho(U_{aj}^{(\rho)} \mid \mathcal{B}_a).$$

This holds true in particular for $a = ()$; according to $\zeta_{()}^{(\rho)} \equiv \rho$ this leads to

$$\lim_{m \to \infty} B^m(\rho) = B(\rho), \ \lim_{m \to \infty} B_j^m(\rho) = B_j(\rho) \quad \forall j \in M.$$

Theorem (4.17) now yields the assertion. $\quad\square$

For the especially important case

$$c(\vartheta, n) \geq \gamma \cdot n \,, \ \gamma > 0,$$

i.e. at least linearly increasing sampling costs, theorem (4.20) together with (2.15) and (2.10) yields the following very intuitive result: At "decision time" $a \in A$ let ξ_a be the

126

a posteriori-distribution and $\varphi_a^{(\xi)}$ be a Bayesian terminal decision function. Defining a sampling plan τ^* by stopping the observation as soon as

$$\psi(\zeta_a(x)) \leq \inf_{j \in M} B_j(\zeta_a(x))$$

(and making a terminal decision according to $\varphi_a^{(\xi)}$) and by taking a further (sub-)sample of size

$$j^* := \min\{k \in M_a : B_k(\zeta_a(x)) = \inf_{j \in M_a} B_j(\zeta_a(x))\}$$

otherwise (because of $\gamma \cdot n \xrightarrow{n \to \infty} \infty$ there exists a k such that the infimum is realized by k) yields a Bayes-optimal procedure $(\tau^*, \varphi^{(\xi)})$.

This result on the one hand represents an analogy with the proposition on p. 122 (for the finite horizon case) and, on the other hand, considerably generalizes the corresponding results for purely sequential procedures (see e.g. [Fe], Ch. 7.2).

V. Optimal sequentially planned tests under side conditions

§ 1   Decision problems with side conditions

From classical decision theory it is well known that Bayes-optimal procedures can, on the one hand, be very important for individual/private decisions and, on the other hand, be of interest for deriving structural properties of „reasonable" procedures. But due to the (at least partial) arbitrariness in determining the prior distribution, these procedures are not accepted by administrations and public health offices. Here requirements concerning the error probabilities (p-values), side conditions concerning the consumer's or producer's risk, restrictions on the total sample size, the sampling costs, the number of stages etc. are of importance. This means that the class of „admissible" decision procedures is restricted by (external) side conditions.

The fundamental Neyman-Pearson lemma already shows that it is, in this case, worth to consider randomized procedures – these make it possible to exhaust completely the given restrictions. Therefore, we will consider randomized sequentially planned statistical decision procedures as introduced in (1.25); the set of all these procedures will be denoted by Δ.

(5.1) Remarks ([Lü], (0.7))

a) Let $\delta = (\beta, \varphi) \in \Delta$ and $\vartheta \in \Theta$. Then

$$P_\vartheta^{(\delta)}(\{a\} \times E) := \int_{\mathcal{X}} b_a^\beta(x, 0) \, \varphi_a(x, E) \, dP_\vartheta(x)$$

defines a probability measure $P_\vartheta^{(\delta)}$ on $(A, \mathcal{P}(A)) \otimes (D, \mathcal{D})$ such that $P_\vartheta^{(\delta)}(A \times D) = 1$.

b) For each $P_\vartheta^{(\delta)}$-integrable function $h : A \times D \to \mathrm{IR}$ follows

$$\int h(a, e) \, dP_\vartheta^{(\delta)} = \sum_{a \in A} \int_{\mathcal{X}} b_a^\beta(x, 0) \int_D h(a, e) \, \varphi_a(x, de) \, dP_\vartheta(x).$$

c) For each randomized sampling plan β and each $\vartheta \in \Theta$

$$P_\vartheta^{(\beta)}(\{a\}) := \int_{\mathcal{X}} b_a^\beta(x, 0) \, dP_\vartheta(x)$$

defines a probability measure on $(A, \mathcal{P}(A))$.

By admitting randomized procedures we gain convexity properties:

(5.2) Remarks (Lü (1.1.3))

a) The set Δ is „convex" i.e. for $\delta' = (\beta', \varphi')$, $\delta'' = (\beta'', \varphi'') \in \Delta$ and $\lambda \in (0; 1)$ there exists a $\delta = (\beta, \varphi) \in \Delta$ such that

$$P_\vartheta^{(\delta)} = \lambda \, P_\vartheta^{(\delta')} + (1 - \lambda) \, P_\vartheta^{(\delta'')} \quad \forall \vartheta \in \Theta.$$

128

b) Let Γ be a ,,convex" subset of Δ and

$$k_\vartheta : (A, \mathcal{P}(A)) \times (D, \mathcal{D}) \to (\mathrm{IR}^n, \mathcal{B}^n), \ \vartheta \in \Theta.$$

Then

$$\{(E_{P_\vartheta^{(\delta)}}(k_\vartheta))_{\vartheta \in \Theta} : \delta \in \Gamma \ \text{s.t. all expectations exist}\}$$

is a convex subset of $(\mathrm{IR}^n)^\Theta$.

Proof: Let $\beta' = (\beta'_a)_{a \in A}$ and $\beta'' = (\beta''_a)_{a \in A}$; then define $\beta = (\beta_a)_{a \in A}$ by

$$\beta_{()} := \lambda \, \beta'_{()} + (1 - \lambda)\beta''_{()}$$

and (inductively) for $a = (a_1, \ldots, a_n), n \in \mathrm{IN}, k \in M \cup \{0\}, \beta_a$ by

$$\beta_a(x, k) := \begin{cases} \beta'_a(x, k) & \text{if } \gamma'_a(x) + \gamma''_a(x) = 0 \\[2mm] \dfrac{\lambda \, b_a^{\beta'}(x, k) + (1 - \lambda)b_a^{\beta''}(x, k)}{\lambda \, \gamma'_a(x) + (1 - \lambda)\gamma''_a(x)} & \text{elsewhere} \end{cases}$$

where $\gamma_a^{'(')}(x) := \prod_{i=0}^{n-1} \beta_{(a_1,\ldots,a_i)}^{'(')}(x, a_{i+1})$; furthermore, let $\varphi = (\varphi_a)_{a \in A}$ be defined by

$$\varphi_a(x, E) := \begin{cases} \varphi'_a(x, E) & \text{if } b_a(x, 0) = 0 \\[2mm] \dfrac{\lambda \, b'_a(x, 0)\varphi'_a(x, E) + (1 - \lambda)b''_a(x, 0)\varphi''_a(x, E)}{b_a(x, 0)} & \text{elsewhere} \end{cases}$$

Then $\delta = (\beta, \varphi) \in \Delta$ and $P_\vartheta^{(\delta)} = \lambda \, P_\vartheta^{(\delta')} + (1 - \lambda)P_\vartheta^{(\delta'')}$. Part b) follows from the linearity of expectations. $\qquad\square$

At the beginning, we mentioned side conditions concerning the error probabilities and the total sampling costs. For randomized tests $\delta = (\beta, \varphi) \in \Delta$ these quantities may be written in the form

$$\begin{aligned} \alpha(\vartheta, \delta) &= P_\vartheta^{(\beta,\varphi)}(A \times \{d_2\}) = \sum_{a \in A} \int_\mathcal{X} b_a^\beta(x, 0) \, \varphi_a(x) \, dP_\vartheta(x) \\ \beta(\vartheta, \delta) &= P_\vartheta^{(\beta,\varphi)}(A \times \{d_1\}) = \sum_{a \in A} \int_\mathcal{X} b_a^\beta(x, 0)(1 - \varphi_a(x)) \, dP_\vartheta(x) \end{aligned}$$

where $\varphi_a(x) := \varphi_a(x, \{d_2\})$

$$K(\vartheta, \delta) = \int C(a) \, dP_\vartheta^{(\beta,\varphi)}(a, e) = \sum_{a \in A} \int_\mathcal{X} b_a^\beta(x, 0) \, C(a) \, dP_\vartheta(x).$$

These terms occur in the following examples which will be treated in detail later on.

(5.3) Example (modified Kiefer-Weiss problem)
Let $\Theta = \{\vartheta_0, \vartheta_1, \vartheta_2\}$ and $\alpha^\star, \beta^\star \in [0, 1]$. We look for a sequentially planned test δ which minimizes $K(\vartheta_0, \delta)$ under (side-)conditions

$$\alpha(\vartheta_1, \delta) \leq \alpha^\star, \quad \beta(\vartheta_2, \delta) \leq \beta^\star. \qquad\qquad\square$$

(5.4) Example (locally optimal sequentially planned tests)

Let $\Theta \subset \mathbb{R}^1$ be open, $\vartheta_0 \in \Theta$, $\alpha^* \in [0,1]$, $K^* \in \mathbb{R}^1_+$, and Γ be the set of all sequentially planned tests whose power function $PF_\delta(\vartheta) := \alpha(\vartheta,\delta)$ is differentiable at ϑ_0. We look for a $\delta \in \Gamma$ which maximizes the derivative $\frac{d}{d\vartheta}\alpha(\vartheta,\delta)|_{\vartheta_0}$ under the (side-) conditions

$$\alpha(\vartheta_0,\delta) = \alpha^* \, , \quad K(\vartheta_0,\delta) \le K^* . \qquad\qquad \square$$

(5.5) Example (weakly admissible tests; Θ finite)

Let $\Theta = \{\vartheta_1,\ldots,\vartheta_n\} = H_1 + H_2$ and

$$L(\vartheta,\delta) := \begin{cases} \alpha(\vartheta,\delta) & \vartheta \in H_1 \\[2mm] & \text{for} \\[2mm] \beta(\vartheta,\delta) & \vartheta \in H_2 . \end{cases}$$

We are interested in tests $\delta \in \Delta$ such that there does not exists any $\delta' \in \Delta$ with

$$\begin{aligned} L(\vartheta_i,\delta') &\le L(\vartheta_i,\delta) \quad \forall 1 \le i \le n, \\ K(\vartheta_i,\delta') &\le K(\vartheta_i,\delta) \quad \forall 1 \le i \le n, \end{aligned}$$

and at least one $<$-relation. $\qquad\qquad \square$

To formulate problems of this kind in a unified way we use for $x = (x_1,\ldots,x_n), y = (y_1,\ldots,y_n) \in \mathbb{R}^n$ the (usual) notation

$$\begin{aligned} x \le y \; &:\Leftrightarrow \; x_i \le y_i \quad \forall 1 \le i \le n \\ x < y \; &:\Leftrightarrow \; x \le y \text{ and } x \ne y . \end{aligned}$$

Let now $\Gamma \subset \Delta$, $r : \Gamma \to \mathbb{R}^n$ be a vector of „characteristics", and $\mathcal{R} := r(\Gamma)$. We look for procedures $\delta^* \in \Gamma$ leading, under certain side conditions, to pareto-minimal values of a function $f : \mathbb{R}^n \to \mathbb{R}^{n_1}$ on [36] $\mathcal{R}$. Let the side conditions be given in the form

$$g\big(r(\delta)\big) \le g^*$$

where $g : \mathbb{R}^n \to \mathbb{R}^{n_2}$ and $g^* \in \mathbb{R}^{n_2}$ (conditions $g(\cdot) = g^*$ will, as usual, be described by $g(\,) \le g^*$ and $g(\cdot) \le -g^*$). Without loss of generality we assume that the set

$$\mathcal{Z} := \{\delta \in \Gamma : g(r(\delta)) \le g^*\}$$

of „admissible" sequentially planned procedures is non empty.

The problem mentioned above may then be formulated as

(5.6) Determine a $\delta^* \in \mathcal{Z}$ such that $\nexists \, \delta \in \mathcal{Z} : f(r(\delta)) < f(r(\delta^*))$.

In the case of a real-valued goal function f this is equivalent with

$$f(r(\delta^*)) \le f(r(\delta)) \qquad \forall \delta \in \mathcal{Z} .$$

Our examples (5.3)-(5.5) fit into this concept in an obvious way:

---

[36]Hence infinite-dimensional goal-functions are excluded.

(i) Modified Kiefer-Weiss problem ((5.3)):
$$r(\delta) := (\alpha(\vartheta_1,\delta),\ \beta(\vartheta_2,\delta),\ K(\vartheta_0,\delta)) \in \mathbb{R}^3$$
$$f(r(\delta)) := K(\vartheta_0,\delta) \in \mathbb{R}^1$$
$$g(r(\delta)) := (\alpha(\vartheta_1,\delta),\beta(\vartheta_2,\delta)) \in \mathbb{R}^2$$
$$g^\star := (\alpha^\star,\beta^\star).$$

(ii) Locally optimal sequentially planned tests ((5.4)):
$$r(\delta) := (\alpha(\vartheta_0,\delta),\alpha(\vartheta_0,\delta);\ \tfrac{d}{d\vartheta}\alpha(\vartheta,\delta)|_{\vartheta_0},\ K(\vartheta_0,\delta)) \in \mathbb{R}^4$$
$$f(r(\delta)) := -\tfrac{d}{d\vartheta}\alpha(\vartheta,\delta)|_{\vartheta_0} \in \mathbb{R}^1$$
$$g(r(\delta)) := (\alpha(\vartheta_0,\delta),\ -\alpha(\vartheta_0,\delta),\ K(\vartheta_0,\delta)) \in \mathbb{R}^3.$$
$$g^\star := (\alpha^\star,-\alpha^\star,K^\star).$$

(iii) Weakly admissible sequentially planned tests ((5.5)):
$$r(\delta) := (L(\vartheta_1,\delta),\ldots,L(\vartheta_n,\delta),\ K(\vartheta_1,\delta),\ldots,K(\vartheta_n,\delta))$$
$$f(r(\delta)) = r(\delta) \in \mathbb{R}^{2n}$$
$$g(r(\delta)) := r(\delta) \in \mathbb{R}^{2n}$$
$$g^\star = (\infty,\ldots,\infty)\ (\text{or } n_2 = 0 \text{ resp.}).$$

§ 2   Characterizations of optimal sequentially planned decision procedures

If $(g,f)(\mathcal{R}) \subset \mathbb{R}^{n_1+n_2}$ is convex and there exist solutions of (5.6), these solutions may also be characterized by using „Lagrangean multipliers":

(5.7) Theorem

Let $(g,f)(\mathcal{R})$ be convex and $\delta^\star$ a solution of (5.6). Then there exists a $\lambda \in \mathbb{R}^{n_1+n_2}$, $\lambda > 0$, such that

$$(\star) \qquad \min_{\delta\in\Gamma}\langle \lambda, (g,f)\circ r(\delta)\rangle = \langle \lambda, (g,f)\circ r(\delta^\star)\rangle.$$

Proof: Let $x^\star := (g,f)\circ r(\delta^\star)$. Then

$$R := (g,f)(\mathcal{R}) = \{(g,f)\circ r(\delta) : \delta \in \Gamma\}$$

is convex (by assumption) as well as

$$Q := \{x = (x_1,\ldots,x_{n_1+n_2}) \in \mathbb{R}^{n_1+n_2} : x_i < x_i^\star\ \forall i\}.$$

Moreover, $R \cap Q = \emptyset$ since the existence of an $\hat{x} = (g,f)\circ r(\hat{\delta}) \in R \cap Q$ would lead to

$$g(r(\hat{\delta})) < g(r(\delta^\star)) \le g^\star\ ,\text{i.e.}\ \hat{\delta} \in \mathcal{Z},\ f(r(\hat{\delta})) < f(r(\delta^\star)),$$

i.e. to a contradiction to the optimality of $\delta^\star$. Hence there exists, according to the separating hyperplane theorem, a $\lambda \in \mathbb{R}^{n_1+n_2}\setminus\{0\}$ such that

$$\langle \lambda, q\rangle \le \langle \lambda, r\rangle \qquad \forall q \in Q, r \in R.$$

By considering $q \nearrow x^\star$ one obtains

$$\langle \lambda, (g,f)\circ r(\delta^\star)\rangle \le \langle \lambda, (g,f)\circ r(\delta)\rangle \qquad \forall \delta \in \Gamma.$$

Finally, since Q is in all components unbounded from below it follows that $\lambda_i \geq 0 \ \forall i.$ $\square$

It is also possible to prove a partial converse of (5.7):

(5.8) Remark

Let $\delta^\star \in \Gamma$ and $\lambda \in \mathrm{IR}^{n_1 + n_2}, \lambda > 0$, fulfil $(\star)$. If

$$(i) \quad \lambda_i > 0 \quad \forall 1 \leq i \leq n_1 + n_2$$

or

$$(ii) \quad r(\delta^\star) \text{ is the unique solution of } (\star),$$

then $\delta^\star$ is a solution of (5.6) for $g^\star = g(r(\delta^\star))$, and

$$\mathcal{Z} = \{\delta \in \Gamma : g(r(\delta)) \leq g(r(\delta^\star))\}.$$

(5.9) Remark

The convexity assumption on $(g, f)(\mathcal{R})$ (in (5.7)) is, in particular, fulfilled if $\mathcal{R}$ is convex and (g, f) is a linear function.

This was the reason to admit randomized procedures; (5.2)b) then guarantees the convexity of $\mathcal{R}$ for a variety of interesting examples; in particular for $\Theta = \{\vartheta_1, \ldots, \vartheta_n\}$

$$\{(L(\vartheta_1, \delta), \ldots, L(\vartheta_n, \delta), K(\vartheta_1, \delta), \ldots, K(\vartheta_n, \delta)) : \delta \in \Delta, \text{ expectations exist}\}$$

is convex (see example (5.5)), and for $\Theta = \{\vartheta_0, \vartheta_1, \vartheta_2\}$

$$\{(\alpha(\vartheta_1, \delta), \ \beta(\vartheta_2, \delta), \ K(\vartheta_0, \delta)) : \delta \in \Delta, \text{ expectations exist}\}$$

is convex (see example (5.3)).

Under the assumption

$$c(\vartheta, n) > 0 \qquad \forall \vartheta \in \Theta, \ n \in \mathrm{IN}$$

(which merely excludes the case of free information) we therefore obtain for our examples (5.3)-(5.5):

(5.10) Examples

a) Let $\delta^\star$ be a solution of the modified Kiefer-Weiss problem (see example (5.3)). According to (5.7) there exists a vector $\lambda = (\lambda_0, \lambda_1, \lambda_2) \neq 0, \lambda_i \geq 0 \ \forall i$, such that $\delta^\star$ minimizes the function

$$\delta \mapsto \lambda_1 \, \alpha(\vartheta_1, \delta) + \lambda_2 \, \beta(\vartheta_2, \delta) + \lambda_0 \, K(\vartheta_0, \delta).$$

If $\alpha(\vartheta_1, \delta^\star), \beta(\vartheta_2, \delta^\star) \in (0, 1)$ and $K(\vartheta_0, \delta^\star) > 0$ then it follows for each $\mathcal{A}$ s.t. $() \in \mathcal{A}$ that $\lambda_1, \lambda_2 > 0$; moreover, if there exists a test $\tilde{\delta}$ such that $(\alpha(\vartheta_1, \tilde{\delta}), \beta(\vartheta_2, \tilde{\delta})) < (\alpha(\vartheta_1, \delta^\star), \beta(\vartheta_2, \delta^\star))$ (which certainly holds for $\mathcal{A} = A$ in the iid case) then $\lambda_0 > 0$. In § 4 we will derive, for special sets $\mathcal{A}$, the structure of tests which solve this problem in the iid case.

b) Let δ^* be a locally optimal test (see example (5.4)); according to (5.7) there exist $\lambda_1, \lambda_3 \geq 0$ and $\lambda_2 \in \mathbb{R}$ (due to the side condition $= \alpha^*$) such that δ^* minimizes the function

$$\delta \mapsto -\lambda_1 \frac{d}{d\vartheta}\alpha(\vartheta,\delta)|_{\vartheta_0} + \lambda_2 \alpha(\vartheta_0,\delta) + \lambda_3\, K(\vartheta_0,\delta).$$

For $\frac{d}{d\vartheta}\alpha(\vartheta,\delta^*)|_{\vartheta_0} \in (0,\infty)$, $\alpha(\vartheta_0,\delta^*) \in (0,1)$, $K(\vartheta_0,\delta^*) > 0$ one obtains in the iid case additionally that

$$\begin{aligned}
\lambda_1 &> 0 \quad \text{for each } \mathcal{A} \text{ s.t. } () \in \mathcal{A}\\
\lambda_3 &> 0 \quad \text{if there exists a test } \tilde{\delta} \text{ s.t.}\\
&\qquad \tfrac{d}{d\vartheta}\alpha(\vartheta,\tilde{\delta})|_{\vartheta_0} > \tfrac{d}{d\vartheta}\alpha(\vartheta,\delta^*)|_{\vartheta_0}, \alpha(\vartheta_0,\tilde{\delta}) = \alpha(\vartheta_0,\delta^*).
\end{aligned}$$

Proof: For $\lambda_1 = 0$ and $\lambda_2 \geq 0$ accepting H_1 without any observation would lead to a contradiction, analogously for $\lambda_1 = 0$ and $\lambda_2 \leq 0$ accepting H_2 without any observation; for $\lambda_3 = 0$ the test $\tilde{\delta}$ would cause a contradiction. Problems of this kind will be treated in § 5.

c) Let δ^* be a weakly admissible test (see example (5.5); $f = id$, $n_2 = 0$). According to (5.7) there exists $\lambda = (\lambda_1, \ldots, \lambda_{2n}) \neq 0$, $\lambda_i \geq 0\ \forall i$, such that δ^* minimizes the function

$$\delta \mapsto \sum_{i=1}^{n}(\lambda_i\, L(\vartheta_i,\delta) + \lambda_{n+1} K(\vartheta_i,\delta))$$

(here $\Gamma = \{\delta \in \Delta$: all expectations exist).

For the special case of testing two simple hypotheses we additionally obtain: If $\alpha(\vartheta_1,\delta^*), \beta(\vartheta_2,\delta^*) \in (0,1)$, and $K(\vartheta_i,\delta^*) > 0$, $i = 1,2$, then for arbitrary $\mathcal{A}$ s.t. $() \in \mathcal{A}$ it follows that $\lambda_1, \lambda_2 > 0$; moreover, if there exists a test $\tilde{\delta}$ such that

$$(\alpha(\vartheta_1,\tilde{\delta}),\ \beta(\vartheta_2,\tilde{\delta})) < (\alpha(\vartheta_1,\delta^*),\ \beta(\vartheta_2,\delta^*))$$

(which certainly holds for $\mathcal{A} = A$ in the iid case) then $\lambda_3 + \lambda_4 > 0$.

d) Let $\Theta = \{\vartheta_1,\vartheta_2\}$, $\mu_1,\mu_2 \geq 0$, $\mu_1 + \mu_2 > 0$, $\alpha^*,\beta^* \in (0,1)$. We look for tests which minimize

$$\mu_1\, K(\vartheta_1,\delta) + \mu_2\, K(\vartheta_2,\delta)$$

under the side conditions

$$\alpha(\vartheta_1,\delta) \leq \alpha^*\,,\ \beta(\vartheta_2,\delta) \leq \beta^*.$$

For each solution δ^* of this problem there exists a vector $\lambda = (\lambda_1,\lambda_2,\lambda_3) \neq 0$, $\lambda_i \geq 0\ \forall i$, such that δ^* minimizes the function

$$\delta \mapsto \lambda_1\, \alpha(\vartheta_1,\delta) + \lambda_2\, \beta(\vartheta_2,\delta) + \lambda_3(\mu_1\, K(\vartheta_1,\delta) + \mu_2\, K(\vartheta_2,\delta));$$

for the iid-case we obtain (analogously to c)) additional properties. $\qquad\square$

Moreover, (5.7) yields assertions on the structure of optimal tests also for the case that the vector of "characteristics" incorporates the expected number of stages, the expected total number of observations, etc.; then one has to minimize terms of the form

$$\sum_{i=1}^{2}(\lambda_1^{(i)}L(\vartheta_i,\delta) + \lambda_2^{(i)}K(\vartheta_i,\delta) + \lambda_3^{(i)}H(\vartheta_i,\delta) + \lambda_4^{(i)}G(\vartheta_i,\delta) + \ldots)$$

where

$$H(\vartheta,\delta) \; := \; \int h(a)dP_\vartheta^{(\delta)}(a,e) \qquad \text{(see (1.21)c))}$$

$$G(\vartheta,\delta) \; := \; \int g(a)dP_\vartheta^{(\delta)}(a,e) \qquad \text{(see (1.21)c))} \qquad \square$$

§ 3 Sequentially planned tests for simple hypotheses in the iid case

In classical theory of testing statistical hypotheses, as considered e.g. by Lehmann [Le], the first step for the derivation/construction of "optimal" (uniformly most powerful, UMP unbiased, UMP invariant,...) tests always consists in considering simple hypotheses $H_i = \{\vartheta_i\}$, $i = 1,2$. Moreover, for simple hypotheses on independently repeated observations the SPRT turns out to be optimal in the sense of Wald and Wolfowitz ([W/W]). Consequently, this special case deserves interest also for sequentially planned procedures. We consider, therefore, the simple hypothese $H_i = \{\vartheta_i\}, i = 1,2$, in the iid case, and we additionally assume that

$$c(\vartheta_1,j) = c(\vartheta_2,j) =: c(j) > 0 \qquad \forall j \in M \text{ and } \lim_{j\to\infty} c(j) = \infty.$$

Evidently, one should restrict attention to weakly admissible tests (see (5.5)), i.e. $\delta \in \Delta$ with the property that there does not exist any $\delta' \in \Delta$ such that

$$\alpha(\vartheta_1,\delta') \leq \alpha(\vartheta_1,\delta), \qquad \beta(\vartheta_2,\delta') \leq \beta(\vartheta_2,\delta),$$
$$K(\vartheta_i,\delta') \leq K(\vartheta_i,\delta), \qquad i = 1,2$$

and at least one "$<$"-relation. In (5.10)c) we described tests of this kind by using "Lagrangean multipliers" λ_i.

Therefore, we now consider for fixed $\lambda_1,\ldots,\lambda_4$, where $\lambda_3 + \lambda_4 > 0$, and $P_{\vartheta_1}^{X_1} \sim P_{\vartheta_2}^{X_1}$ (according to assumption (A); see p. 79) the problem of minimizing

$$(\star) \qquad \delta \mapsto \sum_{i=1}^{2}(\lambda_i\, L(\vartheta_i,\delta) + \lambda_{2+i}\, K(\vartheta_i,\delta)).$$

It turns out that there are strong connections with Bayes-procedures as investigated in Ch. IV.

134

(5.11) Remark

A sequentially planned test $\delta^\star \in \Delta$ minimizes $(\star)$ iff $\delta^\star$ is a Bayes-optimal procedure for

$$L_a(\vartheta_i, d_j, x) \;=\; \begin{cases} \lambda_i & j \neq i \\ & \text{for} \\ 0 & j = i \end{cases} \;,\; 1 \leq i,j \leq 2, a \in A, x \in \mathcal{X},$$

$$c(\vartheta_i, j) \;=\; \lambda_{2+i}\, c(j), \; i = 1,2, \; j \in M,$$

$$\xi \;:=\; \xi(\{\vartheta_1\}) = \xi(\{\vartheta_2\}) = 1/2.$$

Proof: For the corresponding Bayes-problem one has to minimize

$$\frac{1}{2}[\lambda_1\alpha(\vartheta_1, \delta) + \lambda_3 K(\vartheta_1, \delta)] + \frac{1}{2}[\lambda_2\beta(\vartheta_2, \delta) + \lambda_4 K(\vartheta_2, \delta)]. \qquad \square$$

Obviously, this is a special case of the situation (4.12)/(4.16); hence one may apply the results of Ch. IV, § 4. In particular, one obtains optimal tests (i.e. $\delta^\star \in \Delta$ which minimize $(\star)$) by the stopping-/continuation sets

$$B := \{\eta \geq v\}, \; B_j := \{\eta < v, \; j = \inf\{i : v_i = \max_{k \in M} v_k\}\}$$

(see p. 74).

(5.12) Remark

There exist $\pi_1, \pi_2 \in (0,1)$, $\pi_1 \leq \pi_2$ such that

$$B = [0, \pi_1] \cup [\pi_2, 1].$$

Proof: According to our assumptions we obtain from Ch. II, § 2 (see p. 74) that $B = \{\eta \geq v\}$ where η consists (in our case of simple hypotheses) of two linear parts with $\eta(0) = \eta(1) = 0$, and v is convex (and, therefore, continuous on $(0,1)$) with $v \leq -c(1)$; this yields the assertion (see also [Ir], (3.1.7)). $\qquad \square$

At least for concave cost functions c, the following heuristics seem to be reasonable: The smaller the distance of the a posteriori probability (which is strictly isotonic function of the likelihood ratio; see [Ir], p. 81) from the stopping bounds π_i, the smaller is the number of additional observations needed to reach the boundary. Therefore, it seems to be "optimal" to choose smaller additional sample sizes in the neighbourhood of the boundaries π_i than in the middle part of the continuation region, i.e. it should be sufficient to consider sequentially planned tests of the following typ:

(5.13) Definition

a) *For $a \in A$ let $0 < k_1^{(a)} \leq k_2^{(a)} < \infty$ and $\hat{t}_a : \mathrm{IR} \to M_0$ be a measurable function such that*

$$\hat{t}_a(x) \begin{cases} = 0 & \notin \\ & \text{for } x \qquad (k_1^{(a)}, k_2^{(a)}). \\ > 0 & \in \end{cases}$$

The decision procedure (τ, φ) defined by

$$
\begin{aligned}
\tau_1 &:= \hat{t}_{()}(q_{()}(\vartheta_2, \vartheta_1)), \\
\tau_{n+1} &:= \hat{t}_{(\tau_1,\ldots,\tau_n)}(q_{(\tau_1,\ldots,\tau_n)}(\vartheta_2, \vartheta_1)), \; n \in \mathbb{N}, \\
\tau &:= \left(\tau_1, \ldots, \tau_{\min\{j:\tau_j=0\}-1}\right), \\
\varphi_a &:= \begin{cases} 0 & < k_2^{(a)} \\ \quad\; for \; q_a(\vartheta_2, \vartheta_1) \\ 1 & \geq k_2^{(a)} \end{cases}
\end{aligned}
$$

is called <u>generalized sequentially planned probability ratio test</u> (GSPPRT).

b) A sequentially planned test (τ, φ) has <u>onion skins structure</u> if (τ, φ) is a GSPPRT with the property that, for each $a \in A$ and $m \in M$, $\sum_{j \geq m} t_a^{-1}(\{j\})$ is an interval.

In the same way as the GSPRT's of Kiefer and Weiss [K/W] generalize Wald's SPRT's, these GSPPRT's are generalizations of the SPPRT's (see (3.5)) allowing that the stopping bounds as well as the continuation regions may depend on the respective "states" $a \in A$.

The notion onion skins structure explains itself by the graphical representation in figure 5.1

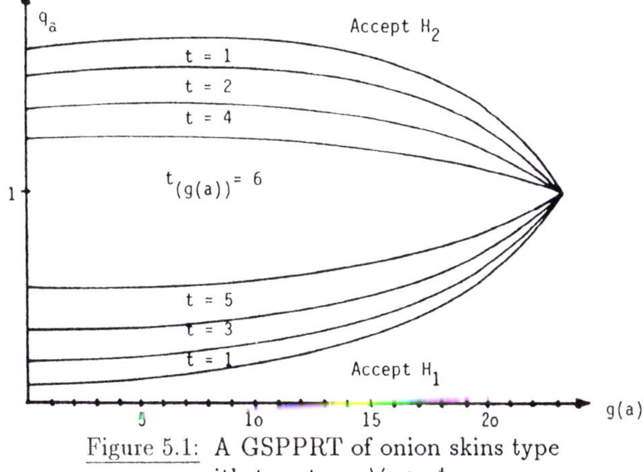

Figure 5.1: A GSPPRT of onion skins type
with $t_a = t_{(g(a))} \; \forall a \in A$

In figure 3.1 already an SPRT with onion skins structure was described. The heuristics mentioned above may now be formulated as

(5.14) "Onion skins conjecture"[37]

In the iid case (with infinite horizon and concave sampling costs) "optimal" sequentially planned tests have onion skins structure, i.e. $\bigcup_{j \geq m} B_j$ is, for each $m \in M$, an interval.

---

[37]N. Schmitz: Wald-Wolfowitz optimality of sequentially planned tests – remarks and conjectures. Ann. of Oper. Res. 32 (1991), p. 213.

136

Indeed, in many cases (also for the optimality concepts treated in § 4 and § 5; see (5.28), (5.42)) optimal tests turn out to be of this structure.

Unfortunately, the following example which is essentially based on an idea of Rieder and Wentges [R/W] shows that this conjecture fails to be true in general.

(5.15) Example
Consider the iid case with $P_\vartheta^{X_1} = \mathcal{B}(1,\vartheta)$, $\vartheta_1 = 0$, $\vartheta_2 = 1/2$, $\lambda_1 = \lambda_2 = 40$ (i.e. $L_a(\vartheta_i, d_j, x) = 40\ \forall i \neq j$, $a \in A_{\mathrm{IN}}$, $x \in \mathcal{X}$), $c(j) = 1 + 2j$, $j \in \mathrm{IN}$, and $\xi = 1/4$.
Then one can easily check that

$$\frac{1}{4} \in B_1, \quad \frac{2}{5} \in B_2, \quad \frac{4}{7} \in B_1, \quad \frac{8}{11} \in B_2, \quad \frac{16}{19} \in B_1, \quad \frac{32}{35} \in B$$

(and, in general, $\xi_a^{(\xi)} \in \{0\} + \{\frac{1}{1+3\cdot 2^{-i}}, i \in \mathrm{IN}\}$).

Therefore, the corresponding Bayes-optimal test (which simultaneously minimizes $(\star)$) fails to have onion skins structure. $\qquad\square$

Since the Bayes-optimality of SPRT's is an essential tool for proving the famous theorem of Wald and Wolfowitz [W/W], it is not too surprising that the Wald-Wolfowitz theorem has no direct counterpart for sequentially planned tests:

(5.16) Example ([Lü], pp. 133-135; footnote 37)
Consider the iid case with $P_\vartheta^{X_1} = \mathcal{B}(1,\vartheta)$, $\vartheta_1 = 0.8$, $\vartheta_2 = 0.2$ and

$$c(j) = \begin{cases} 0.1 + 0.2\,j & \text{for}\quad 1 \leq j \leq 6 \\ 10^6 \cdot j & \text{for}\quad j > 6. \end{cases}$$

Then the SPPRT (τ^*, φ^*) given by

$$\hat{t}^*(x) := \begin{cases} 0 & \leq 4^{-2} \text{ or } \geq 4^3 \\ 2 & = 4^2 \\ 3 & \text{for } x \quad = 4 \\ 4 & = 1 \\ 5 & = 4^{-1} \end{cases}$$

has the error probabilities

$$\alpha^* := \alpha(\vartheta_1, (\tau^*, \varphi^*)) = 0.003777\ldots, \quad \beta^* := \beta(\vartheta_2, (\tau^*, \varphi^*)) = 0.033077\ldots$$

and the expected sampling costs

$$K(\vartheta_1, (\tau^*, \varphi^*)) = 1.117374\ldots, \quad K(\vartheta_2, (\tau^*, \varphi^*)) = 1.46074\ldots$$

(τ^*, φ^*) minimizes $K(\vartheta_1, \delta)$ under the side conditions $\alpha(\vartheta_1, \delta) \leq \alpha^*$, $\beta(\vartheta_2, \delta) \leq \beta^*$. Moreover, (τ^*, φ^*) is a Bayes-optimal sequentially planned test for $\lambda_1 = 2.1 \cdot 10^6$, $\lambda_2 = 2$, $\xi = 65537^{-1}$.

However, the SPPRT (τ', φ') given by

$$\hat{t}'(x) := \begin{cases} 0 & \leq 4^{-2} \text{ or } \geq 4^3 \\ 5 & = 4^2 \\ 6 & \text{for } x \quad = 4 \\ 4 & = 1 \\ 2 & = 4^{-1} \end{cases}$$

has the error probabilities

$$\alpha(\vartheta_1, (\tau', \varphi')) = 0.003772\ldots < \alpha^\star, \ \beta(\vartheta_2, (\tau', \varphi')) = 0.0325\ldots < \beta^\star$$

and the expected sampling costs

$$K(\vartheta_1, (\tau', \varphi')) = 1.11705\ldots < K(\vartheta_1, (\tau^\star, \varphi^\star))$$

(but of course $K(\vartheta_2, (\tau', \varphi')) = 1.6381\ldots > K(\vartheta_2, (\tau^\star, \varphi^\star))$) $\qquad \square$

On the other hand, Lübbert ([Lü], p. 136) has shown that one at a time sampling is not characterized by the Wald-Wolfowitz optimum property.

§ 4   The modified Kiefer-Weiss problem in the iid case

We consider the modified Kiefer-Weiss problem (see (5.3)) in the iid case. Under the weak assumptions of (5.10)a) there exist, for each solution $\delta^\star$ of this problem, Lagrangian multipliers $\lambda_0, \lambda_1, \lambda_2 > 0$ such that $\delta^\star$ minimizes the function

$$\delta \mapsto \lambda_1 \, \alpha(\vartheta_1, \delta) + \lambda_2 \, \beta(\vartheta_2, \delta) + \lambda_0 \, K(\vartheta_0, \delta).$$

Therefore we now consider this problem for fixed λ_i, i.e. are looking for the structure of tests $\delta = (\beta, \varphi)$ which minimize, for $\lambda_1, \lambda_2 > 0$, the term

$$\lambda_1 \, \alpha(\vartheta_1, \delta) + \lambda_2 \, \beta(\vartheta_2, \delta) + K(\vartheta_0, \delta)$$

(according to (5.8) each test of this kind solves a modified Kiefer-Weiss problem)

Using the notation

$$\begin{aligned} L_a^{(i)} &:= q_a(\vartheta_i, \vartheta_0), \ i = 1, 2, \ a \in A \\ c(n) &:= c(\vartheta_0, n), \ n \in \text{IN} \\ C(a) &:= C(\vartheta_0, a), \ a \in A \end{aligned}$$

this minimization problem leads to a problem of optimal sampling (see Ch. II): Since

$$\begin{aligned} &\lambda_1 \, \alpha(\vartheta_1, \delta) + \lambda_2 \, \beta(\vartheta_2, \delta) + K(\vartheta_0, \delta) \\ &= E_{\vartheta_0} \left(\sum_{a \in A} b_a^\beta(\cdot, 0)(\lambda_1 \varphi_a L_a^{(1)} + \lambda_2(1 - \varphi_a)L_a^{(2)} + C(a)) \right) \\ &\geq E_{\vartheta_0} \left(\sum_{a \in A} b_a^\beta(\cdot, 0)(\min\{\lambda_1 L_a^{(1)}, \lambda_2 L_a^{(2)}\} + C(a)) \right) \end{aligned}$$

138

with equality for $\varphi_a = 1_{\{\lambda_1 L_a^{(1)} \leq \lambda_2 L_a^{(2)}\}}$, one has to solve the problem of optimal sampling given by

$$Z_a^{(\lambda_1,\lambda_2)} := E_{\vartheta_0} \left(\sum_{a \in A} \vartheta_a^\beta(\cdot,0)(-\min\{\lambda_1 L_a^{(1)}, \lambda_2 L_a^{(2)}\} - C(a)) \right).$$

Since randomized sampling plans may be considered as non-randomized sampling plans with respect to an enlarged family of σ-algebras (see [Si]; [C/R/S], Theorem 5.3; [Lü] (1.4.12)) where the corresponding essential suprema remain unchanged, we may restrict attention to non-randomized plans. Then the pay-off simplifies to

$$Z_a^{(\lambda_1,\lambda_2)} = -\min\{\lambda_1 L_a^{(1)}, \lambda_2 L_a^{(2)}\} - C(a).$$

In the iid-case $(\lambda_1 L_a^{(1)}, \lambda_2 L_a^{(2)})$ forms a stationary Markov process (on $[0,\infty)^2$) with initial value $\lambda = (\lambda_1, \lambda_2)$. Obviously

$$E_{\vartheta_0}(\sup_{a \in A}(Z_a^{(\lambda)})^+) < \infty;$$

moreover, we obtain

(5.17) Lemma

$$U_a := \operatorname*{ess\,sup}_{\tau \in T_a} E_{\vartheta_0}(Z_\tau^{(\lambda)}|\mathcal{B}_a) = \lim_{m \to \infty} \operatorname*{ess\,sup}_{\substack{\tau \in T_a \\ g(\tau) \leq m}} E_{\vartheta_0}(Z_\tau^{(\lambda)}|\mathcal{B}_a) =: U_a', \ a \in A.$$

Proof: According to (4.18) it is sufficient to show that

$$\lim_{m \to \infty} \int_{\{g(\tau)>m\}} U'_{\tau^{(m)}} \, dP_{\vartheta_0} = 0 \qquad \forall \tau \in T_0;$$

due to $0 \geq U_a' \geq Z_a^{(\lambda)}$ this can, analogously to (4.19), be reduced to

$$\lim_{m \to \infty} \int_{\{g(\tau)>m\}} Z_{\tau^{(m)}}^{(\lambda)} \, dP_{\vartheta_0} = 0.$$

For the second part of $Z_a^{(\lambda)}$ we obtain

$$\begin{aligned}
0 &\leq \int_{\{g(\tau)>m\}} C(\tau^{(m)}) \, dP_{\vartheta_0} \\
&\leq \int_{\{g(\tau)>m\}} C(\tau) \, dP_{\vartheta_0} \ \text{ since } \ \tau \geq \tau^{(m)} \ \text{ and } C \ \text{isotonic} \\
&\leq \int_{\{g(\tau)>m\}} (-Z_\tau^{(\lambda)}) \, dP_{\vartheta_0} \xrightarrow[m \to \infty]{} 0.
\end{aligned}$$

For the first part of $Z_a^{(\lambda)}$ follows

$$\begin{aligned}
0 &\leq \int_{\{g(\tau)>m\}} \min\{\lambda_1 L_{\tau^{(m)}}^{(1)}, \lambda_2 L_{\tau^{(m)}}^{(2)}\} \, dP_{\vartheta_0} \\
&= \lambda_1 \int 1_{\{g(\tau)>m\}} \varphi_{\tau^{(m)}} dP_{\vartheta_1} + \lambda_2 \int 1_{\{g(\tau)>m\}}(1 - \varphi_{\tau^{(m)}}) \, dP_{\vartheta_2},
\end{aligned}$$

where according to the definition of φ

$$\varphi_{\tau(m)} = \begin{cases} 1 \\ 0 \end{cases} \quad \text{if } \lambda_1/\lambda_2 \begin{array}{c} \leq \\ > \end{array} L^{(2)}_{\tau(m)}/L^{(1)}_{\tau(m)}.$$

Since

$$\lim_{m \to \infty} L^{(2)}_{\tau(m)}/L^{(1)}_{\tau(m)} = \begin{cases} 0 & P_{\vartheta_1}\text{ -a.s.} \\ \infty & P_{\vartheta_2}\text{ -a.s.} \end{cases} \quad \text{on } \{g(\tau) = \infty\}$$

(according to the strong law of large numbers) we obtain

$$\lim_{m \to \infty} \varphi_{\tau(m)} = 0 \quad P_{\vartheta_1}\text{ - a.s. on } \{g(\tau) = \infty\}$$
$$\lim_{m \to \infty} \varphi_{\tau(m)} = 1 \quad P_{\vartheta_2}\text{ - a.s. on } \{g(\tau) = \infty\};$$

hence the monotone convergence theorem yields the assertion. $\qquad\square$

Therefore, the results of Ch. II § 4 may be applied. Using the notation

$$\begin{aligned}
v_a(\lambda_1, \lambda_2) &:= \sup_{\tau \succeq a} E_{\vartheta_0} Z^{(\lambda_1, \lambda_2)}_\tau, \; v(\lambda_1, \lambda_2) = v_{()}(\lambda_1, \lambda_2) \\
\eta_a(x_1, x_2) &:= -\min\{x_1, x_2\} - C(a), \; \eta(x_1, x_2) := -\min\{x_1, x_2\} \\
B_0 &:= \{(\lambda_1, \lambda_2) \in [0, \infty)^2 : \eta(\lambda_1, \lambda_2) \geq v(\lambda_1, \lambda_2)\} \\
B_j &:= \{(\lambda_1, \lambda_2) \in [0, \infty)^2 : v_{(j)}(\lambda_1, \lambda_2) \geq v(\lambda_1, \lambda_2)\}, \; j \in M,
\end{aligned}$$

we obtain:

(5.18) Theorem

Let $\mathcal{A} = A$ and β be a randomized sampling plan with $P^{(\beta)}_{\vartheta_0}(A) = 1$. β is optimal iff for P_{ϑ_0}- almost all $x \in \mathcal{X}$, all $j \in M_0$ and all $a \in A$ s.t. $b^\beta_a(x, k) > 0$ for a $k \in M_0$ holds

$$\beta_a(x, j) > 0 \Rightarrow (\lambda_1\, L^{(1)}_a(x), \; \lambda_2\, L^{(2)}_a(x)) \in B_j.$$

If

$$\tau^\star_k := \min\{j \in M_0 : (\lambda_1\, L^{(1)}_{(\tau^\star_1, \ldots, \tau^\star_{k-1})}, \lambda_2\, L^{(2)}_{(\tau^\star_1, \ldots, \tau^\star_{k-1})}) \in B_j\}$$

defines a (non-randomized) sampling plan $\tau^\star$ with the property $P_{\vartheta_0}(\tau^\star \in A) = 1$ then $\tau^\star$ is an optimal sampling plan.

Proof: Obviously

$$\eta_{(a,b)}(\lambda_1, \lambda_2) = \eta_b(\lambda_1, \lambda_2) - C(a).$$

Therefore, the assertion for non-randomized sampling plans follows from Ch. II, §4 (see p. 74). The general case is treated in the way mentioned above (enlarging the σ-algebras without changing the essential suprema). $\qquad\square$

For structural assertions concerning optimal sampling plans we need, therefore, informations about the sets B_j.

A simple criterion ensures that the size of the additional samples required by optimal

140

sampling plans is bounded:

(5.19) Lemma
> *Let* $\lim_{n\to\infty} c(n) = \infty$. *Then there exists a* $j_0 \in \mathbb{N}$ *such that* $B_j = \emptyset \ \forall j \geq j_0$; *in particular, the plan* τ^* *from (5.18) is optimal.*

Proof: Choosing, for $\lambda = (\lambda_1, \lambda_2) \in [0, \infty)^2$, $j_0 \in \mathbb{N}$ such that $c(j_0) > \min\{\lambda_1, \lambda_2\}$ we obtain for all $j \geq j_0$

$$v_{()}(\lambda) \geq \eta(\lambda) > -c(j) \geq \sup_{\tau \succeq (j)} E_{\vartheta_0} Z_\tau^{(\lambda)} = v_{(j)}(\lambda). \qquad \square$$

Similar to the purely sequential case (see [Ir] Ch. 4.2), quite detailed structural properties can be derived for B_0:

$$
\begin{aligned}
B_0 &= \{(\lambda_1, \lambda_2) \in [0, \infty)^2 : \min\{\lambda_1, \lambda_2\} \leq \inf_{\tau \succ ()} (-Z_\tau^{(\lambda_1, \lambda_2)})\} \\
&= \{(\lambda_1, \lambda_2) \in [0, \infty)^2 : \hat{v}(\lambda_1, \lambda_2) \geq \lambda_1 \ \text{ or } \ \hat{v}(\lambda_1, \lambda_2) \geq \lambda_2\}
\end{aligned}
$$

where

$$
\begin{aligned}
\hat{v}(\lambda_1, \lambda_2) &:= \inf_{\tau \succ ()} E_{\vartheta_0}(-Z_\tau^{(\lambda_1, \lambda_2)}) \\
&= \inf_{\substack{(\tau, \varphi) \\ \tau \succ ()}} (\lambda_1 \, E_{\vartheta_1} \varphi_\tau + \lambda_2 \, E_{\vartheta_2}(1 - \varphi_\tau) + E_{\vartheta_0} \, C(\tau));
\end{aligned}
$$

hence we have to analyze the functions

$$\lambda_1 \mapsto \hat{v}(\lambda_1, \lambda_2) - \lambda_1 \ \text{ and } \ \lambda_2 \mapsto \hat{v}(\lambda_1, \lambda_2) - \lambda_2.$$

(5.20) Theorem *(see [Ir], 4.2.4)*
> *There exist isotonic, concave, continuous functions*
> $N_i : [0, \infty) \to [0, \infty)$, $i = 1, 2$, *such that*
>
> *(i)* $\operatorname{sign}(\hat{v}(\lambda_1, \lambda_2) - \lambda_1) = \operatorname{sign}(N_1(\lambda_2) - \lambda_1)$
> $\operatorname{sign}(\hat{v}(\lambda_1, \lambda_2) - \lambda_2) = \operatorname{sign}(N_2(\lambda_1) - \lambda_2)$.
>
> *(ii)* $N_1(0) = N_2(0) = c := \min_{j \in M} c(j)$.
>
> *(iii) If* $\vartheta_i \neq \vartheta_0$ *then* N_i *is bounded,* $i = 1, 2$.
>
> *(iv) There exists a* $\lambda_0 \in [0, \infty)$ *such that*
> $\operatorname{sign}(N_1(\lambda) - \lambda) = \operatorname{sign}(\lambda_0 - \lambda)$
> $\phantom{\operatorname{sign}(N_1(\lambda) - \lambda)} = \operatorname{sign}(N_2(\lambda) - \lambda) \ \forall \lambda \in [0, \infty)$
>
> *(v)* $B_0 = \{(\lambda_1, \lambda_2) \in [0, \infty)^2 : N_1(\lambda_2) \geq \lambda_1 \ \text{or} \ N_2(\lambda_1) \geq \lambda_2\}$

Proof: Due to the symmetry it is sufficient to consider the function N_1.
(i) For fixed λ_2 the function

$$\lambda_1 \mapsto \hat{v}(\lambda_1, \lambda_2) - \lambda_1 = \inf_{\substack{(\tau, \varphi) \\ \tau \succ ()}} (\lambda_1(E_{\vartheta_1} \varphi_\tau - 1) + \lambda_2 E_{\vartheta_2}(1 - \varphi_\tau) + E_{\vartheta_0} C(\tau))$$

is the infimum of linear functions; hence it is concave and continuous. For $\lambda_1 = 0$ its value is > 0 (since $\tau \succ (\,)$) and for $\lambda_1 \to \infty$ its values fall below 0. Therefore, it vanishes for exactly one value $N_1(\lambda_2)$. This function $N_1 : [0, \infty) \to [0, \infty)$ obviously fulfills (i). Moreover,

$$\{(\lambda_1, \lambda_2) \in [0, \infty)^2 : N_1(\lambda_2) \geq \lambda_1\}$$
$$= \{(\lambda_1, \lambda_2) \in [0, \infty)^2 : \hat{v}(\lambda_1, \lambda_2) \geq \lambda_1\}$$
$$= \bigcap_{\substack{(\tau, \varphi) \\ \tau \succ (\,)}} \{(\lambda_1, \lambda_2) \in [0, \infty)^2 : \lambda_1(1 - E_{\vartheta_1}\varphi_\tau) \leq \lambda_2 E_{\vartheta_2}(1 - \varphi_\tau) + E_{\vartheta_0}C(\tau)\}$$
$$= \{(\lambda_1, \lambda_2) \in [0, \infty)^2 : \lambda_1 \leq \inf_{\substack{(\tau, \varphi) \\ \tau \succ (\,), E_{\vartheta_1}\varphi_\tau < 1}} \frac{\lambda_2 E_{\vartheta_2}(1 - \varphi_\tau) + E_{\vartheta_0}C(\tau)}{1 - E_{\vartheta_1}\varphi_\tau}\},$$

i.e.

$$N_1(\lambda_2) = \inf_{\substack{(\tau, \varphi) \\ \tau \succ (\,), E_{\vartheta_1}\varphi_\tau < 1}} \frac{1}{(1 - E_{\vartheta_1}\varphi_\tau)}(\lambda_2 E_{\vartheta_2}(1 - \varphi_\tau) + E_{\vartheta_0}C(\tau)).$$

As an infimum of isotonic linear functions N_1 is isotonic, concave and continuous.

(ii) Since $N_1(0)$ is the unique zero of

$$\hat{v}(\lambda_1, 0) - \lambda_1 = \inf_{\substack{(\tau, \varphi) \\ \tau \succ (\,)}} (\lambda_1 E_{\vartheta_1}\varphi_\tau + E_{\vartheta_0}C(\tau)) - \lambda_1$$
$$= \min_{j \in M} c(j) - \lambda_1$$

we obtain $N_1(0) = \min_{j \in M} c(j)$.

(iii) Let $\vartheta_0 \neq \vartheta_1$, i.e. $P_{\vartheta_0}^{X_1} \neq P_{\vartheta_1}^{X_1}$. Defining for $m := \min M$ a (purely) sequential sampling plan σ by

$$\sigma = \underbrace{(m, \ldots, m)}_{k\text{-times}} \quad \text{if } k = \inf\{n \in \mathrm{IN} : \log L_{m\,n}^{(1)} < 0\}$$

then it follows according to [Ir], p. 136, or [Lor] resp., that

$$P_{\vartheta_0}(v \in A) = 1 \quad \text{and} \quad E_{\vartheta_0}(h(\sigma)) < \infty,$$

and therefore

$$E_{\vartheta_0} C(\sigma) = c(m) E_{\vartheta_0}(h(\sigma)) < \infty \quad \text{and} \quad E_{\vartheta_0}(L_\sigma^{(1)}) < 1.$$

This yields

$$\hat{v}(\lambda_1, \lambda_2) - \lambda_1 \leq \lambda_1 E_{\vartheta_0}(L_\sigma^{(1)}) + C(\sigma) - \lambda_1$$
$$\leq 0 \quad \text{for} \quad \lambda_1 \geq E_{\vartheta_0}C(\sigma)/(1 - E_{\vartheta_0}(L_\sigma^{(1)}))$$

i.e. $N_1(\lambda_2) \leq E_{\vartheta_0}C(\sigma)/(1 - E_{\vartheta_0}(L_\sigma^{(1)}))$. This bound is independent of λ_2; hence N_1 is bounded.

142

(iv) follows from

$$\text{sign}\,(N_1(\lambda) - \lambda) = \text{sign}\,(\hat{v}(\lambda, \lambda) - \lambda) = \text{sign}\,(N_2(\lambda) - \lambda).$$

(v) follows from (i) and the representation

$$B_0 = \{(\lambda_1, \lambda_2) \in [0, \infty)^2 : \hat{v}(\lambda_1, \lambda_2) \geq \lambda_1 \;\; \text{or} \;\; \hat{v}(\lambda_1, \lambda_2) \geq \lambda_2\}. \qquad \square$$

Theorem (5.20) allows a rather detailed description of the geometric structure of B_0 or B_0^c resp. (see figure 5.2).

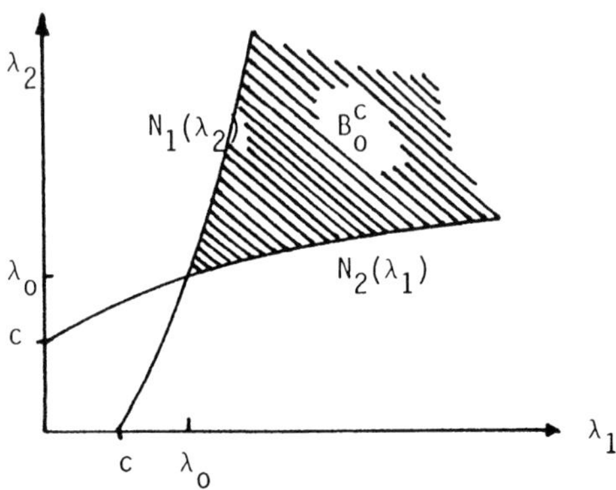

Figure 5.2: Geometric representation of B_0^c

Analogously to the purely sequential case we can also prove a structural assertion concerning the "crossing point" λ_0:

(5.21)Lemma

$$\min_{j \in M} c(j) \leq \lambda_0 \leq \inf_{j \in M} c(j) \frac{1}{1 - P_{\vartheta_1}(L_{(j)}^{(1)} > L_{(j)}^{(2)}) - P_{\vartheta_2}(L_{(j)}^{(1)} \leq L_{(j)}^{(2)})};$$

if $\lim_{n \to \infty} c(n) = \infty$ *and* $\min\{L_j^{(1)}, L_j^{(2)}\} \leq 1 \; \forall j \in M$ *then the second inequality is sharp.*

Proof: According to (5.20) we have $\lambda_0 = \hat{v}(\lambda_0, \lambda_0)$. From $\hat{v}(\lambda_0, \lambda_0) \geq \min_{j \in M} c(j)$ follows the first inequality, and from

$$\hat{v}(\lambda_0, \lambda_0) \;=\; \inf_{\tau \succ ()} E_{\vartheta_0}(C(\tau) + \lambda_0 \min\{L_\tau^{(1)}, L_\tau^{(2)}\})$$

$$\leq\; c(j) + \lambda_0 E_{\vartheta_0} \min\{L_{(j)}^{(1)}, L_{(j)}^{(2)}\} \quad \forall j \in M$$

we obtain the second inequality. Under the additional assumptions follows on the one hand

$$\min\,\{\lambda_0 L_{(j)}^{(1)}, \lambda_0 L_{(j)}^{(2)}\} \leq \lambda_0 \qquad \forall j \in M$$

and therefore

$$(\lambda_0\, L^{(1)}_{(j)},\; \lambda_0\, L^{(2)}_{(j)}) \in B_0 \qquad \forall j \in M;$$

this yields

$$\inf_{\tau \succ ()} E_{\vartheta_0}(C(\tau)\; +\; \lambda_0\, \min\{L^{(1)}_\tau, L^{(2)}_\tau\}) =$$
$$= \inf_{j \in M}(c(j) + \lambda_0\, E_{\vartheta_0}(\min\{L^{(1)}_{(j)}, L^{(2)}_{(j)}\})).$$

On the other hand there exists a $j_0 \in M$ such that the last infimum is attained; hence the assertion follows. $\qquad\qquad\square$

Due to the monotonicity of N_i the limit $\lim_{\lambda \to \infty} N_i(\lambda)$ exist (in $[0, \infty]$), $i = 1, 2$. In the purely sequential case these limits may be computed (see [Lor], [Ir], 4.2.4 (iii)), but in the general case the influence of the cost function prevents a corresponding result.

For the special (but quite important) case of an one-parameter exponential family with densities of the form

$$f_\vartheta(x) = h(x)\, \exp(\vartheta x - \eta(\vartheta))$$

and parameter values $\vartheta_1 < \vartheta_0 < \vartheta_2$, the structure of optimal sampling plans may also be described by using the likelihood ratios $q_a(\vartheta_2, \vartheta_1)$: According to [Ir], p. 141, there exist $s, t \in \mathrm{IR}$ and a bijective mapping $T : (0, \infty)^2 \to \mathrm{IR}^2$ (independent of λ_1, λ_2) such that

$$T(\lambda_1 L^{(1)}_a,\; \lambda_2\, L^{(2)}_a) = (s + \log(q_a(\vartheta_2, \vartheta_1)), t - g(a)) \quad \forall a \in A$$

and

$$T(\{(x, y) \in (0, \infty)^2 : x \le y\}) = \{(x, y) \in \mathrm{IR}^2 : x \ge 0\}.$$

Because of

$$(\lambda_1 L^{(1)}_a, \lambda_2 L^{(2)}_a) \in B \Leftrightarrow (s + \log(q_a(\vartheta_1, \vartheta_2)), t - g(a)) \in T(B)$$

we therefore obtain on the one hand for the terminal decisions

$$\varphi_a = 1 \;\;\Leftrightarrow\;\; (\lambda_1 L^{(1)}_a, \lambda_2 L^{(2)}_a) \in \{(x, y) \in (0, \infty)^2 : x \le y\}$$
$$\Leftrightarrow\;\; s + \log q_a(\vartheta_2, \vartheta_1) \ge 0,$$

and on the other hand, in complete analogy to [Ir] 4.2.6, for the set $T(B_0)$:

(5.22) Lemma *(see [Lor])*

Let a one-parameter exponential family (in natural parametrization be given. Then there exist functions $\tilde{N}_1, \tilde{N}_2 : \mathrm{IR} \to \mathrm{IR}$ with the properties

(i) $\tilde{N}_1$ is continuous and isotonic; $\lim_{y \to \infty} \tilde{N}_1(y) = -\infty$
$\tilde{N}_2$ is continuous and antitonic; $\lim_{y \to -\infty} \tilde{N}_2(y) = \infty$,

(ii) sign $\tilde{N}_1(x) = -$ sign $\tilde{N}_2(x) = $ sign $(x - r)$ for a suitable $r \in \mathrm{IR}$

such that

$$T(B_0) = \{(x, y) \in \mathrm{IR}^2 : x \le \tilde{N}_2(y)\ \text{or}\ x \ge \tilde{N}_1(y)\}.$$

144

(5.22) provides detailed information about the geometric structure of $T(B_0)$; see figure 5.3

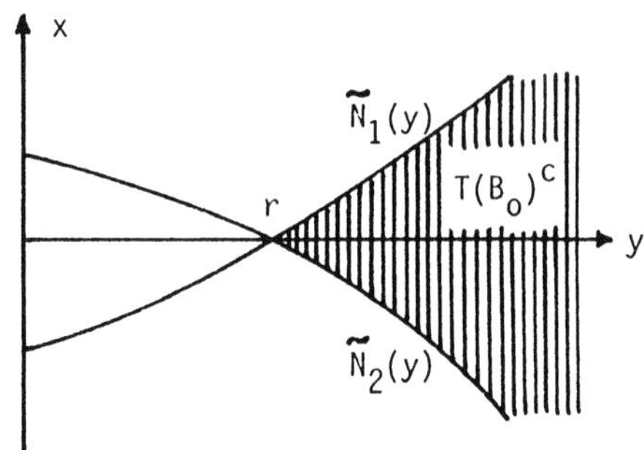

Figure 5.3: Geometric structure of $T(B_0)^c$

This knowledge together with (5.19) and (5.18) yields the following assertion:

(5.23) Theorem

Let a one-parameter exponential family (in natural parametrization) be given, $\vartheta_1 < \vartheta_0 < \vartheta_2$, $\mathcal{A} = A$, $\lim_{n\to\infty} c(n) = \infty$, and $\lambda_1, \lambda_2 > 0$. Then there exists a GSPPRT (with respect to ϑ_1, ϑ_2) with the properties

(i) $\hat{t}_a = \hat{t}_{(g(a))} \quad \forall a \in A$,

(ii) $n \to k_1^{(n)}$ is isotonic; $n \to k_2^{(n)}$ is antitonic,

(iii) there exists an $n_0 \in \mathbb{N}$ such that

$$k_1^{(a)} = k_2^{(a)} \, \forall a \in A : g(a) \geq n_0,$$

(iv) there exists an $s \in \mathbb{R}$ such that

$$k_1^{(a)} \leq s \leq k_2^{(a)} \, \forall a \in A,$$

which minimizes

$$\lambda_1 \, \alpha(\vartheta_1, \delta) + \lambda_2 \, \beta(\vartheta_2, \delta) + K(\vartheta_0, \delta)$$

(and, therefore, solves the corresponding modified Kiefer-Weiss-problem).

This shows once more that GSPPRT's are not only formal generalizations of previous procedures but also solutions of sensible optimization problems.

Until now we considered the case $\mathcal{A} = A$ (infinite horizon); next we treat the finite horizon case $|\mathcal{A}| < \infty$ where optimal sampling plans may be constructed by backward induction (see Ch. II, §2; §4). Using similar notation as before, in particular

$$v_{a,j}(\lambda_1, \lambda_2) \;:=\; \sup_{\substack{\tau \succeq (j) \\ (a,\tau) \in \mathcal{A}}} E_{\vartheta_0}\, \eta_{(a,\tau)}((\lambda_1 L^{(1)}_\tau, \lambda_2 L^{(2)}_\tau))$$

$$v^\star_a(\lambda_1, \lambda_2) \;:=\; \sup_{j \in M_0} v_{a,j}(\lambda_1, \lambda_2)$$

$$B_{a,j} \;:=\; \{(\lambda_1, \lambda_2) \in [0, \infty)^2 : v_{a,j}(\lambda_1, \lambda_2) \geq v^\star_a(\lambda_1, \lambda_2)\},$$

we obtain from (2.26)

(5.24) Theorem

Let $|\mathcal{A}| < \infty$. Then a randomized sampling plan β is optimal iff for P_{ϑ_0}-almost all $x \in \mathcal{X}$, all $j \in M_0$ and all $a \in A$ s.t. $b^\beta_a(x, k) > 0$ for a $k \in M_0$ holds

$$\beta_a(x, j) > 0 \Rightarrow (\lambda_1 L^{(1)}_a(x), \lambda_2 L^{(2)}_a(x)) \in B_{a,j}.$$

τ^* *defined by*

$$\tau^*_k := \min\{j \in M_0 : (\lambda_1 L^{(1)}_{(\tau^\star_1, \ldots, \tau^\star_{k-1})}, \lambda_2 L^{(2)}_{(\tau^\star_1, \ldots, \tau^\star_{k-1})}) \in B_{(\tau^\star_1, \ldots, \tau^\star_{k-1}), j}\}$$

is a (non-randomized) optimal sampling plan.

In particular

$$\begin{aligned}
B_{a,0} &= \{(\lambda_1, \lambda_2) \in [0, \infty) : -\min\{\lambda_1, \lambda_2\} - C(a) \geq v^\star_a(\lambda_1, \lambda_2)\} \\
&= \{(\lambda_1, \lambda_2) \in [0, \infty) : \lambda_1 \leq \hat{v}_a(\lambda_1, \lambda_2) \text{ or } \lambda_2 \leq \hat{v}_a(\lambda_1, \lambda_2)\}
\end{aligned}$$

where

$$\hat{v}_a(\lambda_1, \lambda_2) := \inf_{\substack{\tau > () \\ (a,\tau) \in \mathcal{A}}} (\lambda_1 E_{\vartheta_1}\varphi_\tau + \lambda_2 E_{\vartheta_2}(1 - \varphi_\tau) + C(\tau)).$$

Being an infimum of linear functions, $\hat{v}_a(\lambda_1, \lambda_2) - \lambda_1$ is concave and continuous; hence $\{(\lambda_1, \lambda_2) : \lambda_1 \leq \hat{v}_a(\lambda_1, \lambda_2)\}$ is a closed, convex subset of $[0, \infty)^2$. Furthermore,

$$\hat{v}_a(\lambda_1, \lambda_2) \geq c := \min_{j \in M} c(j)$$

and therefore

$$\begin{aligned}
\{(\lambda_1, \lambda_2) : \lambda_1 \leq c\} &\subset \{(\lambda_1, \lambda_2) : \lambda_1 \leq \hat{v}_a(\lambda_1, \lambda_2)\} \subset B_{a,0}, \\
\{(\lambda_1, \lambda_2) : \lambda_1 < c\} &\subset B_{a,0} \cap \bigcap_{j \in M} (B_{a,j})^c.
\end{aligned}$$

For $x \in \{(\lambda_1, \lambda_2) : \lambda_1 \leq \hat{v}_a(\lambda_1, \lambda_2)\}$ we obtain by convexity- and continuity-arguments

$$\{(\lambda_1, \lambda_2) : \lambda_1 \leq x_1, \lambda_2 \geq x_2\} \subset \{(\lambda_1, \lambda_2) : \lambda_1 \leq \hat{v}_a(\lambda_1, \lambda_2)\} \subset B_{a,0}.$$

Analogous assertions hold for $\hat{v}_a(\lambda_1, \lambda_2) - \lambda_2$ etc.; this leads to

146

(5.25) Theorem

Let a one-parameter exponential family (in natural parametrization) be given, $\vartheta_1 < \vartheta_0 < \vartheta_2$, and $\lambda_1, \lambda_2 > 0$. Then there exists a GSPPRT (with respect to ϑ_1, ϑ_2) which minimizes

$$\lambda_1 \, \alpha(\vartheta_1, \delta) + \lambda_2 \, \beta(\vartheta_2, \delta) + K(\vartheta_0, \delta)$$

(and, therefore, solves the corresponding modified Kiefer-Weiss-problem).

Proof: From

$$
\begin{aligned}
L_a(x) &= q_a(\vartheta_2, \vartheta_1)(x) = \exp(\underbrace{(\vartheta_2 - \vartheta_1)}_{>0} \sum_{i=1}^{g(a)} x_i - (\eta(\vartheta_2) - \eta(\vartheta_1))g(a)), \\[2ex]
L_a^{(1)} &= q_a(\vartheta_1, \vartheta_0)(x) = \exp(\underbrace{(\vartheta_1 - \vartheta_0)}_{<0} \sum_{i=1}^{g(a)} x_i - (\eta(\vartheta_1) - \eta(\vartheta_0))g(a)), \\[2ex]
L_a^{(2)} &= q_a(\vartheta_2, \vartheta_0)(x) = \exp(\underbrace{(\vartheta_2 - \vartheta_0)}_{>0} \sum_{i=1}^{g(a)} x_i - (\eta(\vartheta_2) - \eta(\vartheta_0))g(a)),
\end{aligned}
$$

one obtains

$$L_a(x) \le L_a(x') \Rightarrow \left\{ \begin{array}{l} L_a^{(1)}(x) \ge L_a^{(1)}(x') \\ L_a^{(2)}(x) \le L_a^{(2)}(x'). \end{array} \right.$$

Assume now that there exist x, x', x'' such that $L_a(x) < L_a(x') < L_a(x'')$ and

$$(\lambda_1 L_a^{(1)}(x), \lambda_2 L_a^{(2)}(x)), (\lambda_1 L_a^{(1)}(x''), \lambda_2 L_a^{(2)}(x'')) \in (B_{a,0})^c$$
$$(\lambda_1 L_a^{(1)}(x'), \lambda_2 L_a^{(2)}(x')) \in B_{a,0},$$

where w.l.o.g.

$$(\lambda_1 L_a^{(1)}(x'), \lambda_2 L_a^{(2)}(x')) \in \{(\lambda_1, \lambda_2) : \lambda_1 \le \hat{v}(\lambda_1, \lambda_2)\}.$$

But on the other hand

$$L_a^{(1)}(x'') \le L_a^{(1)}(x'), \quad L_a^{(2)}(x'') \ge L_a^{(2)}(x'')$$

which yields the contradiction

$$(\lambda_1 L_a^{(1)}(x''), \ \lambda_2 L_a^{(2)}(x'')) \in B_{a,0};$$

hence that situation cannot occur.

Furthermore,

$$
\begin{aligned}
L_a(x) \to \infty &\Rightarrow (\lambda_1 L_a^{(1)}(x), \ \lambda_2 L_a^{(2)}(x)) \to (0, \infty) \\
L_a(x) \to 0 &\Rightarrow (\lambda_1 L_a^{(1)}(x), \ \lambda_2 L_a^{(2)}(x)) \to (\infty, 0);
\end{aligned}
$$

i.e.

$$(\lambda_1 L_a^{(1)}(x), \ \lambda_2 L_a^{(2)}(x)) \in B_{a,0} \cap \bigcap_{j \in M} (B_{aj})^c$$

for $L_a(x)$ sufficiently large (and analogously for $L_a(x)$ sufficiently small). Therefore, there exist $k_1^{(a)}, k_2^{(a)} \in (0, \infty)$ such that

$$L_a(x) \in [k_1^{(a)}, k_2^{(a)}]^c \Rightarrow (\lambda_1 L_a^{(1)}(x),\ \lambda_2 L_a^{(2)}(x)) \in B_{a,0} \cap \bigcap_{j \in M} (B_{aj})^c.$$

Choosing $k_1^{(a)}$ maximal and $k_2^{(a)}$ minimal (with that property) yields

$$L_a(x) \in (k_1^{(a)}, k_2^{(a)}) \Rightarrow (\lambda_1 L_a^{(1)}(x), \lambda_2 L_a^{(2)}(x)) \in B_{a,0}^c. \qquad \square$$

(5.26) Remark

Let $k \in \mathrm{IN}$ and $\mathcal{A} := \{a \in A : g(a) \le k\}$. Then

$$v_{a,j}(\lambda_1, \lambda_2) = \sup_{\substack{\tau \succeq (j) \\ g(\tau) \le k - g(a)}} E_{\vartheta_0}(-\min\{\lambda_1 L_\tau^{(1)}, \lambda_2 L_\tau^{(2)}\} - C(\tau)) - C(a),$$

$$v_a^\star(\lambda_1, \lambda_2) = \sup_{g(\tau) \le k - g(a)} E_{\vartheta_0}(-\min\{\lambda_1 L_\tau^{(1)}, \lambda_2 L_\tau^{(2)}\} - C(\tau)) - C(a),$$

i.e. the regions $B_{a,j}$ depend on a over $g(a)$ only. For the special case of an exponential family we, therefore, obtain from (5.25) that there exists an optimal GSPPRT with $\hat{t}_a = \hat{t}_{(g(a))} \; \forall a \in \mathcal{A}$. $\qquad \square$

These results give quite detailed descriptions of optimal sequentially planned tests. For the (numerical) computation of these procedures, the following recursion turns out to be useful: Define

$$Q_j((\lambda_1, \lambda_2), B) := P_{\vartheta_0}((\lambda_1 L_{(j)}^{(1)}, \lambda_2 L_{(j)}^{(2)}) \in B).$$

Then we obtain, due to the Markov structure,

$$v_{a,j}(\lambda_1, \lambda_2) = \int v_{(a,j)}^\star(\lambda_1', \lambda_2') Q_j((\lambda_1, \lambda_2), d(\lambda_1', \lambda_2')),$$

which obviously simplifies the calculations needed for the backward induction.

For an illustration we consider a special kind of exponential families: Let the densities be of the form

$$f_\vartheta(x) = h(x) \exp(\zeta(\vartheta)x - \eta(\vartheta)),$$

leading to

$$q_a(\vartheta', \vartheta) = \exp(\gamma_1(\vartheta, \vartheta') \sum_{n=1}^{g(a)} X_n - \gamma_0(\vartheta, \vartheta')g(a))$$

where

$$\gamma_1(\vartheta, \vartheta') = \zeta(\vartheta') - \zeta(\vartheta), \qquad \gamma_0(\vartheta, \vartheta') = \eta(\vartheta) - \eta(\vartheta'),$$

and assume that the X_n are integer-valued. Since

$$\lambda_i L_a^{(i)} = \lambda_i \exp(\gamma_1(\vartheta_0, \vartheta_i) \sum_{n=1}^{g(a)} X_n - \gamma_0(\vartheta_0, \vartheta_i)g(a))$$

$(i = 1, 2)$, we need, for the construction of optimal tests, only the values of

$$f_{a,j}(i) \; := \; v_{a,j}(\lambda_1 \exp(\gamma_1(\vartheta_0, \vartheta_1)i - \gamma_0(\vartheta_0, \vartheta_1)g(a)), \lambda_2 \exp(\gamma_1(\vartheta_0, \vartheta_2)i - \gamma_0(\vartheta_0, \vartheta_2)g(a)))$$
$$f_a^\star(i) \; := \; v_a^\star(\lambda_1 \exp(\gamma_1(\vartheta_0, \vartheta_1)i - \gamma_0(\vartheta_0, \vartheta_1)g(a)), \lambda_2 \exp(\gamma_1(\vartheta_0, \vartheta_2)i - \gamma_0(\vartheta_0, \vartheta_2)g(a)))$$

for the $i \in \mathrm{supp}(\sum_{n=1}^{g(a)} X_n)$. Applying the recursion formula yields

(5.27) Lemma

Under these assumptions

$$
\begin{aligned}
f_{a,j}(i) &= \sum_{k \in supp(\sum_{n=1}^{j} X_n)} f_{(a,j)}^\star(i + k) \cdot P_{\vartheta_0}(\sum_{n=1}^{j} X_n = k) \; \forall j \in M \\
f_{a,0}(i) &= - \min_{n=1,2} \lambda_n \exp(\gamma_1(\vartheta_0, \vartheta_n)i - \gamma_0(\vartheta_0, \vartheta_n)g(a)) - C(a), \\
f_a^\star(i) &= \sup_{j \in M_0} f_{a,j}(i),
\end{aligned}
$$

and for each "endpoint" $a \in \mathcal{A}$ (i.e. $a \in \mathcal{A}$ s.t. $\nexists b \succ () : (a, b) \in \mathcal{A}$) holds

$$f_a^\star(i) = f_{a,0}(i).$$

According to (5.26), a further simplification is possible in the case $\mathcal{A} = \{a \in A : g(a) \leq k\}$. This is utilized in the following example which is a further variation of our illustrative example on p. 10.

(5.28) Example (see p. 10; (3.34))
Again we consider the case of $\mathcal{B}(1, p)$ (binomial-) distribution with

$$p_1 = 0.90016837105, \quad p_2 = 0.95,$$

and additionally $p_0 = 0.925$. For the SPRT $\delta_{1/9,9}$ we obtain[38]

$$
\begin{aligned}
\alpha(p_1, \delta_{1/9,9}) &= 0.0998, \quad \beta(p_2, \delta_{1/9,9}) = 0.0791 \\
\mathrm{ASN}_{\delta_{1/9,9}}(p_1) &= 96.46, \quad \mathrm{ASN}_{\delta_{1/9,9}}(p_2) = 111.65
\end{aligned}
$$

and

$$\mathrm{ASN}_{\delta_{1/9,9}}(p_0) = 143.72,$$

i.e. we have a considerable improvement comparing with the binomial test with fixed sample size 199 (which guarantees the same accuracy at p_1 and p_2; see p. 11).

But for the cost function $c(n) = 5 + n$ this SPRT yields $\mathrm{ASC}_{\delta_{1/9,9}}(p) = 6 \cdot \mathrm{ASN}_{\delta_{1/9,9}}(p)$, i.e. the disastrously bad values

$$\mathrm{ASC}_{\delta_{1/9,9}}(p_1) = 578.76 \quad \mathrm{ASC}_{\delta_{1/9,9}}(p_2) = 669.90 \quad \mathrm{ASC}_{\delta_{1/9,9}}(p_0) = 862.32,$$

whereas the fixed sample size test leads to sampling costs 204.

---

[38]All these computations are due to Th. Meyerthole.

Using our previous results we may now try to find, by variation of λ_1 and λ_2, a GSPPRT which also keeps the same error probabilities as the SPRT. To simplify calculations we consider only tests which use as group sizes multiples of 25, i.e. $M = 25 \cdot \mathrm{IN}$, and we allow for at most 225 observations, i.e. $\mathcal{A} = \{a \in A_M : g(a) \leq 225\}$. By backward induction one may construct a GSPPRT $\delta^\star$ which minimizes for $\lambda_1 = 900$, $\lambda_2 = 1000$ the term

$$\lambda_1\,\alpha(\vartheta_1,\delta) + \lambda_2\,\beta(\vartheta_2,\delta) + K(\vartheta_0,\delta);$$

this test is given by

$$\hat{t}_{(0)}(\exp(\gamma_1(\vartheta_1,\vartheta_2)i - \gamma_0(\vartheta_1,\vartheta_2)\cdot 0)) \;=\; \begin{cases} 100 & \text{for } i = 0 \\ 0 & \text{elsewhere} \end{cases}$$

$$\hat{t}_{(100)}(\exp(\gamma_1(\vartheta_1,\vartheta_2)i - \gamma_0(\vartheta_1,\vartheta_2)\cdot 100)) \;=\; \begin{cases} 50 & \text{for } i = 92,93,94 \\ 25 & \text{for } i = 91,95 \\ 0 & \text{elsewhere} \end{cases},$$

$$\hat{t}_{(125)}(\exp(\gamma_1(\vartheta_1,\vartheta_2)i - \gamma_0(\vartheta_1,\vartheta_2)\cdot 125)) \;=\; \begin{cases} 50 & \text{for } i = 116,117 \\ 25 & \text{for } i = 114,115,118 \\ 0 & \text{elsewhere} \end{cases},$$

$$\hat{t}_{(150)}(\exp(\gamma_1(\vartheta_1,\vartheta_2)i - \gamma_0(\vartheta_1,\vartheta_2)\cdot 150)) \;=\; \begin{cases} 50 & \text{for } i = 139,140 \\ 25 & \text{for } i = 138,141 \\ 0 & \text{elsewhere} \end{cases},$$

$$\hat{t}_{(175)}(\exp(\gamma_1(\vartheta_1,\vartheta_2)i - \gamma_0(\vartheta_1,\vartheta_2)\cdot 175)) \;=\; \begin{cases} 25 & \text{for } i = 161,\ldots,164 \\ 0 & \text{elsewhere} \end{cases},$$

$$\hat{t}_{(200)}(\exp(\gamma_1(\vartheta_1,\vartheta_2)i - \gamma_0(\vartheta_1,\vartheta_2)\cdot 200)) \;=\; \begin{cases} 25 & \text{for } i = 185,186 \\ 0 & \text{elsewhere} \end{cases},$$

$$\hat{t}_{(225)}(\exp(\gamma_1(\vartheta_1,\vartheta_2)i - \gamma_0(\vartheta_1,\vartheta_2)\cdot 225)) \;\equiv\; 0,$$

and

$$\begin{aligned}
k_2^{(100)} &= \exp(\gamma_1(\vartheta_1,\vartheta_2)\cdot 96 - \gamma_0(\vartheta_1,\vartheta)\cdot 100) \\
k_2^{(125)} &= \exp(\gamma_1(\vartheta_1,\vartheta_2)\cdot 119 - \gamma_0(\vartheta_1,\vartheta_2)\cdot 125) \\
k_2^{(150)} &= \exp(\gamma_1(\vartheta_1,\vartheta_2)\cdot 142 - \gamma_0(\vartheta_1,\vartheta_2)\cdot 150) \\
k_2^{(175)} &= \exp(\gamma_1(\vartheta_1,\vartheta_2)\cdot 165 - \gamma_0(\vartheta_1,\vartheta_2)\cdot 175) \\
k_2^{(200)} &= \exp(\gamma_1(\vartheta_1,\vartheta_2)\cdot 187 - \gamma_0(\vartheta_1,\vartheta_2)\cdot 200) \\
k_2^{(225)} &= \exp(\gamma_1(\vartheta_1,\vartheta_2)\cdot 209 - \gamma_0(\vartheta_1,\vartheta_2)\cdot 225).
\end{aligned}$$

This GSPPRT, which obviously is of the "onion skins"-type, leads to the error probabilities

$$\alpha(\vartheta_1,\delta^\star) = 0.0986\,, \quad \beta(\vartheta_2,\delta^\star) = 0.0786,$$

i.e. $\delta^\star$ has a slightly higher accuracy as the SPRT $\delta_{1/9,9}$ and the fixed sample size test, and to the average sampling costs

$$ASC_{\delta^\star}(\vartheta_1) = 138.76 \quad ASC_{\delta^\star}(\vartheta_0) = 164.63 \quad ASC_{\delta^\star}(\vartheta_2) = 146.08,$$

150

i.e. δ^* improves the SPRT in a dramatical way and is considerably better than the fixed sample size test (and moreover, has the upper bound 225 for the total sample size). According to (5.8) this test possesses the following optimum property: Under the side conditions

$$\alpha(\vartheta_1, \delta) \leq 0.0986, \qquad \beta(\vartheta_2, \delta) \leq 0.0786$$

δ^* minimizes the expected sampling costs $K(\vartheta_0, \delta)$, i.e. δ^* solves the corresponding modified Kiefer-Weiss-problem. $\qquad\qquad\square$

§ 5 Locally optimal sequentially planned tests in the dominated iid case

In this section we consider, for the dominated case, the problem of constructing locally optimal sequentially planned tests (see (5.4)); for the purely sequential case this problem was treated by Berk [Bek] and Irle [Ir]. Concerning the differentiability concept we refer to [Ir], (4.1.3); for the special case of one-parameter exponential families we then obtain

(5.29) Remark ([Ir], (4.1.3))

Let $(Q_\vartheta)_{\vartheta\in\Theta}$ be a one-parameter exponential family with densities of the form

$$f_\vartheta(x) = h(x) \; \exp(\zeta(\vartheta)x - \eta(\vartheta))$$

where ζ and η are differentiable. Then $(Q_\vartheta)_{\vartheta\in\Theta}$ is differentiable for each $\vartheta \in \Theta$. For the derivative r follows

$$r(x) = \frac{d}{d\vartheta} \log f_\vartheta(x)|_{\vartheta=\vartheta_0} = \zeta'(\vartheta_0)x - \eta'(\vartheta_0);$$

moreover,

$$E_{Q_{\vartheta_0}} id = \eta'(\vartheta_0)/\zeta'(\vartheta_0).$$

By using the stopping rule $\sigma := g(\tau)$ one obtains for each sequentially planned test a purely sequential test with the same power function; hence lemma (4.1.4) of [Ir] yields:

(5.30) Remark

Let $(P_\vartheta^{X_1})_{\vartheta\in\Theta}$ be differentiable in ϑ_0 with derivative r, let[39]

$$\limsup_{\vartheta\to\vartheta_0} \frac{I(\vartheta_0, \vartheta)}{(\vartheta - \vartheta_0)^2} < \infty,$$

and let (β, φ) be a randomized sequentially planned test such that $G(\vartheta_0, (\beta, \varphi)) < \infty$. Then the power function of (β, φ) is differentiable at ϑ_0 with derivative

$$E_{\vartheta_0}\Big(\sum_{a\in A} b_a^\beta(\cdot, 0)\varphi_a \sum_{i=1}^{g(a)} r(X_i)\Big).$$

---

[39] $I(\vartheta, \eta)$ denotes the Kullback-Leibler-information; see p.61.

Remark (4.1.8) of [Ir] shows that the assumptions of (5.30) are, in particular, fulfilled for one-parameter exponential families with densities of the form

$$f_\vartheta(x) = h(x) \cdot \exp(\vartheta x - \eta(\vartheta)).$$

Our aim is to characterize the structure of locally optimal sequentially planned tests in the dominated iid-case. Let $\Theta \subset \mathrm{I\!R}$, $\vartheta_0 \in \overset{\circ}{\Theta}$ and δ^* locally optimal, i.e. a solution of

$$\alpha'(\vartheta_0, \delta) \overset{!}{=} \max$$

under the side conditions

$$\alpha(\vartheta_0, \delta) = \alpha^* , \quad K(\vartheta_0, \delta) \leq K^\star.$$

According to (5.10)b) there exist, under the weak conditions given there, real numbers $\lambda_1 > 0$, λ_2 and $\lambda_3 > 0$ such that δ^* minimizes the function

$$\delta \mapsto -\lambda_1\, \alpha'(\vartheta_0, \delta) + \lambda_2\, \alpha(\vartheta_0, \delta) + \lambda_3\, K(\vartheta_0, \delta).$$

Therefore we now consider this problem for fixed λ_i. Under the assumptions of (5.30) holds

$$\alpha'(\vartheta_0, (\beta, \varphi)) = E_{\vartheta_0}\Big(\sum_{a \in A} b_a^\beta(\cdot, 0) \varphi_a\, S_a\Big)$$

where $S_a = \sum_{i=1}^{g(a)} r(X_i)$, $(r(X_i))_{i \in \mathrm{I\!N}}$ iid, and $E_{\vartheta_0} r(X_1) = 0$. Therefore the minimization problem turns out to be equivalent to the problem of maximizing

$$E_{\vartheta_0}\Big(\sum_{a \in A} b_a^\beta(\cdot, 0)\, (\varphi_a\, S_a + \lambda\, \varphi_a - \mu\, C(a))\Big)$$

where $\mu > 0$, $\lambda \in \mathrm{I\!R}$ and $C(a) := C(\vartheta_0, a)$ (as in § 4). For a non-randomized test (τ, φ) this term simplifies to

$$E_{\vartheta_0}(\varphi_\tau S_\tau) + \lambda\, E_{\vartheta_0}\varphi_\tau - \mu\, E_{\vartheta_0} C(\tau).$$

Since

$$E_{\vartheta_0}\Big(\sum_{u \in A} b_a^\beta(\cdot, 0)\, (\varphi_a S_a + \lambda\, \varphi_a - \mu\, C(a))\Big)$$
$$\leq E_{\vartheta_0}\Big(\sum_{a \in A} b_a^\beta(\cdot, 0)((S_a + \lambda)^+ - \mu\, C(a))\Big)$$

where equality holds for $\varphi_a = 1_{\{S_a + \lambda \geq 0\}}$, we again arive at a problem of optimal sampling.

Due to the same arguments as in § 4 we may restrict attention to non-randomized sampling plans.

For the infinite horizon case $\mathcal{A} = A$, $(S_a + \lambda)_{a \in A}$ is a stationary Markov process with initial state λ; hence the results of Ch. II, § 4 are applicable. But these results require (see p. 35) that

$$E_{\vartheta_0}(\sup_{a \in A}(Z_a^{(\lambda)})^+) < \infty$$

152

where $Z_a^{(\lambda)} := (S_a + \lambda)^+ - \mu\, C(a)$. Unfortunately, already the simple example

$$P_\vartheta^{X_1} = \mathcal{N}(\vartheta, \frac{1}{2\pi}), \ \lambda = 0, \ \mu = 1, \ c(n) := c(\vartheta_0, n) = \log n$$

shows that this condition may be violated in our case:

$$
\begin{aligned}
E_{\vartheta_0}(\sup_{a \in A}((S_a + \lambda)^+ \ - \ \mu\, C(a))) &\geq \sup_{n \in \mathbb{N}} E_{\vartheta_0}(S_n^+ - c(n))) \\
&= \sup_{n \in \mathbb{N}}(\sqrt{n} - \log n) = \infty.
\end{aligned}
$$

Therefore, we will *assume* throughout this section that

$$(\star) \qquad E_{\vartheta_0} \sup_{a \in A}((S_a + \lambda)^+ - \mu\, C(a))^+ < \infty. \qquad \forall\, \lambda \in \mathbb{R},\ \mu > 0.$$

Our next assertion shows that this condition is fulfilled in many interesting cases, in particular for

(i) $c(n) = \infty \quad \forall n > n_0$

(i.e. n_0 is an upper bound for the size of the additional sub-samples)

(ii) $c(n) = c_0 + c \cdot n$

(i.e. fixed costs and a linear part):

(5.31) Lemma

>*If there exists a $\gamma > 0$ such that $c(n) \geq \gamma \cdot n \ \forall n \in \mathbb{N}$ and $E_{\vartheta_0} r(X_1)^2 < \infty$ then condition $(\star)$ is fulfilled.*

Proof:

$$
\begin{aligned}
((S_a + \lambda)^+ \ - \ \mu\, C(a))^+ &\leq ((S_a + \lambda)^+ - \mu\gamma g(a))^+ \\
&= (S_a + \lambda - \mu\gamma g(a))^+, \ \text{since } \mu\gamma g(a) \geq 0 \\
&\leq \lambda^+ + (S_a - \mu\gamma g(a))^+, \text{since } (c+d)^+ \leq c^+ + d^+ \ \forall c, d \in \mathbb{R} \\
&\leq \lambda^+ + \sup_{n \in \mathbb{N}}(S_n - \mu\gamma n)^+;
\end{aligned}
$$

therefore it follows from [Ir], p. 170/171, that

$$E_{\vartheta_0}(\sup_{a \in A}((S_a + \lambda)^+ - \mu C(a))^+) \leq \lambda^+ + E_{\vartheta_0} \sup_{n \in \mathbb{N}}(S_n - \mu\gamma n)^+ < \infty. \qquad \square$$

For solving our problem of optimal sampling one obviously may restrict attention to sampling plans τ with $E_{\vartheta_0}(Z_\tau^{(\lambda)})^- < \infty$. The next remark characterizes these sampling plans and, moreover, shows that the class of these plans is independent of λ and μ.

(5.32) Remark

$$E_{\vartheta_0}(Z_\tau^{(\lambda)})^- < \infty \quad \text{iff} \quad E_{\vartheta_0} C(\tau) < \infty.$$

Proof: Using condition $(\star)$ the assertion follows from

$$- \mu C(\tau) \leq -(Z_\tau^{(\lambda)})^- \leq Z_\tau^{(\lambda)} \leq \sup_{a \in A}((S_a + \lambda)^+ - \frac{\mu}{2}C(a))^+ - \frac{\mu}{2}C(\tau). \qquad \square$$

Using the same notation as in Ch. IV, § 4, we may now formulate:

(5.33) Lemma

For $\tau \in T_{()}$ holds

$$\lim_{m \to \infty} \int_{\{g(\tau) > m\}} U'_{\tau(m)} \, dP_{\vartheta_0} = 0$$

and (therefore) $U_a = U'_a \quad \forall a \in A$.

Proof: According to the definition of U^m_a we obtain for all $m \in \mathrm{I\!N}$

$$E_{\vartheta_0}(\sup_{a \in A}(Z^{(\lambda)}_a)^+) \geq U^m_a \geq Z^{(\lambda)}_a = (S_a + \lambda)^+ - \mu C(a) \geq -\mu C(a);$$

this yields

$$E_{\vartheta_0}(\sup_{a \in A}(Z^{(\lambda)}_a)^+) \geq U'_a \geq -\mu C(a).$$

Therefore, it suffices to show that

$$\lim_{m \to \infty} \int_{\{g(\tau) > m\}} E_{\vartheta_0}(\sup_{a \in A}(Z^{(\lambda)}_a)^+) \, dP_{\vartheta_0} = 0$$

and

$$\lim_{m \to \infty} \int_{\{g(\tau) > m\}} C(\tau^{(m)}) \, dP_{\vartheta_0} = 0.$$

The first part follows from $\lim_{m \to \infty} P_{\vartheta_0}(g(\tau) > m) = 0$ and assumption $(\star)$. Moreover,

$$0 \leq \int_{\{g(\tau) > m\}} C(\tau^{(m)}) \, dP_{\vartheta_0} \leq \int_{\{g(\tau) > m\}} C(\tau) \, dP_{\vartheta_0} \xrightarrow[m \to \infty]{} 0$$

since $\lim_{m \to \infty} P_{\vartheta_0}(g(\tau) > m) = 0$ and $E_{\vartheta_0} C(\tau) < \infty$ according to (5.32); this yields the second part. The second assertion of (5.33) follows from (4.18). $\qquad\square$

Therefore, we may now apply the results of Ch. IV. § 4; in particular we obtain in the same way as in § 5 (see (5.18)):

(5.34) Theorem

Let $\mathcal{A} = A$ and β be a randomized sampling plan with $P^{(\beta)}_{\vartheta_0}(A) = 1$. β is optimal iff for P_{ϑ_0}-almost all $x \in \mathcal{X}$, all $j \in M_0$ and all $a \in A$ s.t. $b^\beta_a(x, k)$ > 0 for a $k \in M_0$ holds

$$\beta_a(x, j) > 0 \quad \Rightarrow \quad S_a(x) + \lambda \in B_j.$$

If

$$\tau^\star_k := \min\{j \in M_0 : S_{\tau^\star_1 + \ldots + \tau^\star_{k-1}} + \lambda \in B_j\}$$

defines a (non-randomized) sampling plan $\tau^\star$ with the property $P_{\vartheta_0}(\tau^\star \in A) = 1$ then $\tau^\star$ is an optimal sampling plan.

For structural assertions concerning optimal sampling plans we need, therefore, again informations about the sets B_j. Analogously to (5.19) we obtain

154

(5.35) Lemma

Let $\lim_{n\to\infty} c(n) = \infty$. Then there exists a $j_0 \in \mathrm{I\!N}$ such that $B_j = \emptyset \; \forall j \geq j_0$; in particular, the plan τ^* from (5.34) is optimal.

Proof:

$$
\begin{aligned}
Z_\tau^{(\lambda)} &\leq \left((S_\tau + \lambda)^+ - \frac{\mu}{2}C(\tau)\right)^+ - \frac{\mu}{2}C(\tau) \\
&= \left(S_\tau + \lambda - \frac{\mu}{2}C(\tau)\right)^+ - \frac{\mu}{2}C(\tau) \\
&\leq \lambda^+ + \left(S_\tau - \frac{\mu}{2}C(\tau)\right)^+ - \frac{\mu}{2}C(\tau) \\
&= \lambda^+ + \left(S_\tau^+ - \frac{\mu}{2}C(\tau)\right)^+ - \frac{\mu}{2}C(\tau) \\
&\leq \lambda^+ + \sup_{a \in A}\left(S_a^+ - \frac{\mu}{2}C(a)\right)^+ - \frac{\mu}{2}C(\tau).
\end{aligned}
$$

This yields

$$
\begin{aligned}
v_{(j)}(\lambda) &= \sup_\tau E_{\vartheta_0}\left((S_{(j,\tau)} + \lambda)^+ - \mu(c(j) + C(\tau))\right) \\
&\leq \lambda^+ + E_{\vartheta_0} \sup_{a \in A}\left(S_a^+ - \frac{\mu}{2}C(a)\right)^+ - \frac{\mu}{2}c(j).
\end{aligned}
$$

Now we choose $j_0 \in \mathrm{I\!N}$ such that

$$
c(j) > \frac{2}{\mu} E_{\vartheta_0} \sup_{a \in A}\left(S_a^+ - \frac{\mu}{2}C(a)\right)^+ \qquad \forall j \geq j_0;
$$

the existence of such a j_0 (independent of λ) follows from condition $(\star)$ and $\lim_{n\to\infty} c(n) = \infty$. Hence we obtain

$$
v_{(j)}(\lambda) < \lambda^+ \qquad \forall \lambda \in \mathrm{I\!R}, \; j \geq j_0
$$

and therefore $B_j = \emptyset \;\; \forall j \geq j_0$. $\qquad\square$

To derive structural properties of the sets B_j one needs informations of the functions v_a.

(5.36) Lemma

(i) For each $a \in A$ the function v_a is isotonic and convex (and therefore continuous).

(ii) The function $\lambda \mapsto v(\lambda) - \lambda^+$ is isotonic and convex on $\mathrm{I\!R}_{\leq 0}$ and antitonic and convex on $\mathrm{I\!R}_{\geq 0}$.

Proof: The monotonicity of v_a follows from the fact that

$$
(S_a + \lambda_1)^+ \leq (S_a + \lambda_2)^+ \qquad \forall \lambda_1 < \lambda_2.
$$

Let now $\lambda_1, \lambda_2 \in \mathrm{I\!R}$, $\gamma \in (0; 1)$ and $\tau \succeq ()$. Then

$$
\begin{aligned}
(S_\tau + \gamma\,\lambda_1 + (1-\gamma)\lambda_2)^+ &= (\gamma(S_\tau + \lambda_1) + (1-\gamma)(S_\tau + \lambda_2))^+ \\
&\leq \gamma(S_\tau + \lambda_1)^+ + (1-\gamma)(S_\tau + \lambda_2)^+
\end{aligned}
$$

and therefore

$$E_{\vartheta_0}((S_\tau + \gamma \lambda_1 + (1 - \gamma)\lambda_2)^+ - \mu\, C(\tau))$$
$$\leq \gamma\, E_{\vartheta_0}((S_\tau + \lambda_1)^+ - \mu\, C(\tau)) + (1 - \gamma)E_{\vartheta_0}((S_\tau + \lambda_2)^+ - \mu\, C(\tau)),$$

i.e. the function $\lambda \mapsto E_{\vartheta_0}((S_\tau + \lambda)^+ - \mu\, C(\tau))$ is convex. Since suprema of convex functions are again convex, assertion (i) follows. $v(\lambda) - \lambda^+$ obviously coincides with $v(\lambda)$ on $\mathrm{IR}_{\leq 0}$; this yields the first part of (ii). Since the difference of a convex and a linear function is again convex, $v(\lambda) - \lambda^+$ is convex on $\mathrm{IR}_{\geq 0}$. The antitonicity follows from the monotonicity of

$$(S_a + \lambda)^+ - \lambda = \max\{S_a, -\lambda\}. \qquad\qquad \square$$

As a consequence of (5.36) we obtain some structural properties of B_j^c:

(5.37) Lemma

B_0^c *is an open interval which is either empty or contains 0. For each $k \in \mathrm{IN}$* $(B_o \cup \ldots \cup B_k)^c \subset \mathrm{IR}$ *is open.*

Proof: Because of $B_0^c = \{v - \eta > 0\}$ the first part follows from (5.36); see also figure 5.4. Moreover,

$$(B_0 \cup \ldots \cup B_k)^c = \{v - \eta > 0\} \cap \{v - v_{(1)} > 0\} \cap \ldots \cap \{v - v_{(k)} > 0\};$$

hence the second part results from the continuity of the functions v_a.

$$\square$$

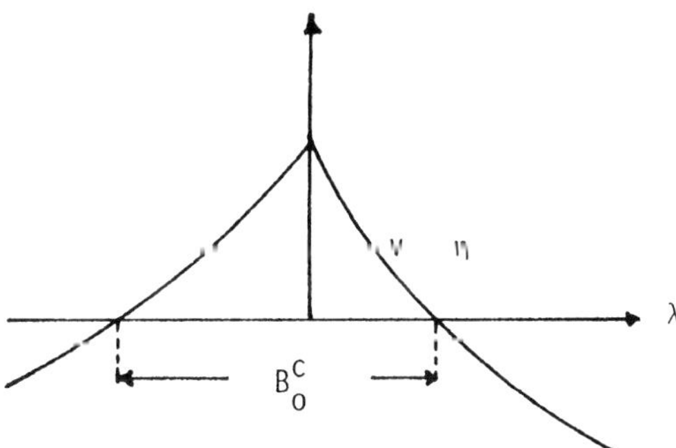

Figure 5.4: Geometric representation of B_0^c

Slightly strengthening the assumption of (5.35) we may ensure the property $P_{\vartheta_0}(\tau^\star \in A) = 1$ and, therefore, the optimality of $\tau^\star$.

156

(5.38) Theorem

> *Assume that there exists a $\gamma > 0$ such that $c(n) \geq \gamma n \ \forall n \in \mathbb{N}$, and that $0 < Var_{\vartheta_0} r(X_1) < \infty$; let τ^* as in (5.34).*
>
> *(i) B_0^c is bounded, in particular*
>
> $$B_0^c \subset \left(-\frac{Var_{\vartheta_0} r(X_1)}{\mu\gamma}, \frac{Var_{\vartheta_0} r(X_1)}{\mu\gamma} \right)$$
>
> *(ii) $P_{\vartheta_0}(\tau^* \in A) = 1$.*
>
> *(iii) τ^* is optimal.*

Proof: Part (i) follows analogously to [Ir], p. 124. Then part (ii) can be shown analogously to p. 68/69; the optimality of τ^* finally follows from theorem (5.34). $\qquad\square$

These results show that locally optimal sequentially planned tests have a similar structure as the SPPRT's – the role of $\log q_a(\vartheta_1, \vartheta_2)$ is now played by $\sum_{i=1}^{g(a)} r(X_i)$. For the special case of one-parameter exponential families and $\vartheta_0 \in \overset{\circ}{\Theta}$ one can, moreover, prove the existence of infinitely many pairs $(\vartheta_1, \vartheta_2) \in \Theta^2$, $\vartheta_1 < \vartheta_0 < \vartheta_2$, such that these tests are indeed SPPRT's with respect to the parameter-values ϑ_1, ϑ_2 (comp. [Ir], p. 128):

(5.39) Theorem

> *Assume that $\mathcal{A} = A$, that a one-parameter exponential family with densities of the form*
>
> $$f_\vartheta(x) = h(x) \ \exp(\vartheta x - \eta(\vartheta))$$
>
> *is given, and that there exists a $\gamma > 0$ such that $c(n) \geq \gamma n \ \forall n \in \mathbb{N}$.*
> *Then there exist infinitely many $(\vartheta_1, \vartheta_2) \in \Theta^2$, $\vartheta_1 < \vartheta_0 < \vartheta_2$, and SPPRT's with respect to ϑ_1, ϑ_2 with bounded $\hat{t}$ which are optimal (i.e. minimize $\delta \mapsto -\lambda_1 \alpha'(\vartheta_0, \delta) + \lambda_2 \ \alpha(\vartheta_0, \delta) + \lambda_3 K(\vartheta_0, \delta))$.*

Proof: According to (5.30) (or [Ir], (4.1.8)) the power function is differentiable. Under the assumption on the exponential family follows that $Var_{\vartheta_0} r(X_i) \in (0, \infty) \ \forall \vartheta_0 \in \overset{\circ}{\Theta}$; therefore, according to lemma (5.31), the condition $(\star)$ is fulfilled. Theorem (5.38) together with the transformation mentioned by Irle ([Ir], p. 128) yields the assertion. $\qquad\square$

This result gives a further justification for the investigations in Chapter III.

Up to now we considered the infinite horizon case $\mathcal{A} = A$. For the case $|\mathcal{A}| < \infty$ optimal sampling plans may be constructed by backward induction (see Ch. II, § 2, § 4). Analogously to (5.24) we obtain from (2.26)

(5.40) Theorem

> *Let $|\mathcal{A}| < \infty$. Then a randomized sampling plan β is optimal iff for P_{ϑ_0}-almost all $x \in \mathcal{X}$, all $j \in M_0$ and all $a \in A$ s.t. $b_a^\beta(x, k) > 0$ for a $k \in M_0$ holds*
>
> $$\beta_a(x, j) > 0 \quad \Rightarrow \quad S_a(x) + \lambda \in B_{a,j}.$$
>
> *τ^* defined by*
>
> $$\tau_k^* := \min\{j \in M_0 : S_{\tau_1^* + \ldots + \tau_{k-1}^*} + \lambda \in B_{(\tau_1^*, \ldots, \tau_{k-1}^*), j}$$
>
> *is a (non-randomized) optimal sampling plan.*

Moreover, one obtains analogously to (5.36)-(5.39)

(5.41) Theorem

> *(i) $(B_{a,0})^c$ is an open interval.*
>
> *(ii) If $0 < Var_{\vartheta_0} r(X_1) < \infty$ then $(B_{a,0})^c$ is an open bounded interval.*
>
> *(iii) Let a one-parameter exponential family (in natural parametrization) be given. Then there exists a GSPPRT which minimizes*
>
> $$-\lambda_1 \alpha'(\vartheta_0, \delta) + \lambda_2 \alpha(\vartheta_0, \delta) + \lambda_3 K(\vartheta_0, \delta).$$

For the special case $\mathcal{A} = \{a \in A : g(a) \le k\}$ it follows (comp. (5.26)) that the regions $B_{a,j}$ depend on a over $g(a)$ only. As a consequence one obtains that for one-parameter exponential families there exists an optimal GSPPRT with $\hat{t}_a = \hat{t}_{(g(a))} \ \forall a \in A$.

For the numerical computation of these procedures the recursive relation

$$v_{a,j}(y) = \int v_{(a,j)}^*(y') \, Q_j(y, dy'),$$

where $Q_j(y, B) := P_{\vartheta_0}(S_j + y \in B)$, turns out to be useful (comp. p. 147). These additional structures are used for the computations[40] carried out for a further modification of our illustrative example from p. 10.

(5.42) Example

For $P_\vartheta^{X_1} = \mathcal{B}(n, p)$ we obtain

$$f_\vartheta(x) = \binom{n}{x} \exp(x \log(\frac{\vartheta}{1-\vartheta}) - n \, \log(\frac{1}{1-\vartheta})),$$

i.e. $\zeta(\vartheta) = \log \frac{\vartheta}{1-\vartheta}$, $\eta(\vartheta) = n \, \log \frac{1}{1-\vartheta}$, and therefore

$$\zeta'(\vartheta) = \frac{1}{\vartheta(1-\vartheta)} \, , \quad \eta'(\vartheta) = n \frac{1}{1-\vartheta}.$$

---

[40] Again all these computations are due to Th. Meyerthole

Again we consider (comp. (5.32))

$$p_1 = 0.90016837105 \,, \quad p_2 = 0.95$$

and $p_0 = 0.925$. Varying the values of μ and λ one can try to find a GSPPRT which guarantees for p_1 and p_2 the same accuracy as the SPRT $\delta_{1/9,9}$ and, moreover, is locally optimal at ϑ_0. Again we restrict attention to $\mathcal{A} = \{a \in A_M : g(a) \leq 225\}$, $M = 25 \cdot \mathrm{IN}$, and the cost function $c(n) = 5 + n$. For $\mu = 0.044$ and $\lambda = -6$ one can construct (by backward induction) a GSPPRT δ^* which minimizes

$$\alpha'(\vartheta_0, \delta) + \lambda \alpha(\vartheta_0, \delta) - \mu K(\vartheta_0, \delta);$$

δ^* is given by

$$\hat{t}_{(0)}(\zeta'(\vartheta_0)i - \eta'(\vartheta_0) \cdot 0) = \begin{cases} 100 & \text{for } i = 0 \\ 0 & \text{elsewhere,} \end{cases}$$

$$\hat{t}_{(100)}(\zeta'(\vartheta_0)i - \eta'(\vartheta_0) \cdot 100) = \begin{cases} 75 & \text{for } i = 93 \\ 50 & \text{for } i = 92, 94 \\ 25 & \text{for } i = 91, 95 \\ 0 & \text{elsewhere,} \end{cases}$$

$$\hat{t}_{(125)}(\zeta'(\vartheta_0)i - \eta'(\vartheta_0) \cdot 125) = \begin{cases} 50 & \text{for } i = 115, 116, 117 \\ 25 & \text{for } i = 114, 118 \\ 0 & \text{elsewhere,} \end{cases}$$

$$\hat{t}_{(150)}(\zeta'(\vartheta_0)i - \eta'(\vartheta_0) \cdot 150) = \begin{cases} 50 & \text{for } i = 139, 140 \\ 25 & \text{for } i = 138, 141 \\ 0 & \text{elsewhere,} \end{cases}$$

$$\hat{t}_{(175)}(\zeta'(\vartheta_0)i - \eta'(\vartheta_0) \cdot 175) = \begin{cases} 50 & \text{for } i = 162, 163 \\ 25 & \text{for } i = 161, 164 \\ 0 & \text{elsewhere,} \end{cases}$$

$$\hat{t}_{(200)}(\zeta'(\vartheta_0)i - \eta'(\vartheta_0) \cdot 200) = \begin{cases} 25 & \text{for } i = 185, 186, 187 \\ 0 & \text{elsewhere,} \end{cases}$$

$$\hat{t}_{(225)}(\zeta'(\vartheta_0)i - \eta'(\vartheta_0) \cdot 225) \equiv 0,$$

and

$$\begin{aligned}
k_2^{(100)} &= (\zeta'(\vartheta_0) \cdot 96 - \eta'(\vartheta_0)100), \\
k_2^{(125)} &= (\zeta'(\vartheta_0) \cdot 119 - \eta'(\vartheta_0)125), \\
k_2^{(150)} &= (\zeta'(\vartheta_0) \cdot 142 - \eta'(\vartheta_0)150), \\
k_2^{(175)} &= (\zeta'(\vartheta_0) \cdot 165 - \eta'(\vartheta_0)175), \\
k_2^{(200)} &= (\zeta'(\vartheta_0) \cdot 188 - \eta'(\vartheta_0)200), \\
k_2^{(225)} &= (\zeta'(\vartheta_0) \cdot 209 - \eta'(\vartheta_0)225).
\end{aligned}$$

This GSPPRT δ^*, which is also of the "onion skins"-type, leads to the error probabilities

$$\alpha(\vartheta_1, \delta^*) = 0.0959 \,, \quad \beta(\vartheta_2, \delta^*) = 0.0779,$$

i.e. it has a slightly higher accuracy than the SPRT $\delta_{1/9,9}$ and the test with fixed sample size 199. Its expected sampling costs are

$$ASC_{\delta^*}(\vartheta_1) = 140.74 \quad ASC_{\delta^*}(\vartheta_0) = 168.06 \quad ASC_{\delta^*}(\vartheta_2) = 148.25,$$

i.e. δ^* is much better than the SPRT and it is considerably better than the fixed sample size test.

Moreover, according to (5.8) this test is optimal in the following sense: Under the side conditions

$$\begin{aligned}
\alpha(\vartheta_0, \delta) &\leq \alpha(\vartheta_0, \delta^*) = 0.4643 \\
K(\vartheta_0, \delta) &\leq K(\vartheta_0, \delta^*) = 168.06
\end{aligned}$$

δ^* maximizes the derivative $\alpha'(\vartheta_0, \delta)$ of the power function at ϑ_0; hence δ^* is locally optimal. One obtains $\alpha'(\vartheta_0, \delta^*) = 21.39$. $\qquad\qquad\square$

§ 6 Remarks on the monotonicity of the power functions of SPPRT's and GSPPRT's

For several optimization problems (under side conditions), SPPRT's and GSPPRT's turned out to be optimal solutions. These optimization problems concerned only very few (≤ 3) parameter values $\vartheta \in \Theta$ or local properties. Since SPPRT's/GSPPRT's may also be used e.g. for one-sided testing problems (where infinitely many parameter values are involved), one needs further informations on the behaviour of the power-/OC-function of these tests.

In the purely sequential case, the power-function of each GSPRT is monotone if the families $(P_\vartheta^{(X_1,\ldots,X_n)})_{\vartheta \in \Theta}$ have monotone likelihood ratio for all $n \in \mathrm{I\!N}$. This property results from the unbiasedness of the GSPRT's (see [Gh], 3.2.2).

Unfortunately, neither the unbiasedness nor the monotonicity of the power-functions of GSPPRT's carry over to the general case (for SPPRT's comp. p. 63).

(5.43) Example
Let $P_\vartheta^{X_1} = \mathcal{B}(1,\vartheta)$, $\Theta = (0;1)$, $\vartheta_1 < \frac{1}{2}$ and $\vartheta_2 = 1 - \vartheta_1$; then

$$q(\vartheta_2, \vartheta_1)(x) = \frac{f_{\vartheta_2}(x)}{f_{\vartheta_1}(x)} = \left(\frac{\vartheta_1}{\vartheta_2}\right)^{1-2x}, \quad x \in \{0,1\}.$$

Define a GSPPRT by

$$\hat{t}_{()} := 1_{(\frac{1}{2},\frac{3}{2})},$$

$$\hat{t}_{(1)}(y) := \begin{cases} 1 & y = \vartheta_1/\vartheta_2 \\ & \text{for} \\ 2 & y = \vartheta_2/\vartheta_1 \end{cases}$$

160

$\hat{t}_{(1,1)}$ given by

$$k_1^{(1,1)} = k_2^{(1,1)} = \min\left\{\left(\frac{\vartheta_1}{\vartheta_2}\right)^2, \left(\frac{\vartheta_2}{\vartheta_1}\right)^2\right\} = \left(\frac{\vartheta_1}{\vartheta_2}\right)^2,$$

and $\hat{t}_{(1,2)}$ given by

$$k_1^{(1,2)} = k_2^{(1,2)} > \left(\frac{\vartheta_2}{\vartheta_1}\right)^3.$$

Then it follows from

$$
\begin{aligned}
P_{\vartheta_1}(\varphi_\tau = 1) &= P_{\vartheta_1}(\hat{t}_{(1)}(q_{(1)}(\vartheta_2,\vartheta_1)) = 1) = P_{\vartheta_1}(X_1 = 0) = \vartheta_2 \\
P_{\vartheta_2}(\varphi_\tau = 0) &= P_{\vartheta_2}(\hat{t}_{(1)}(q_{(1)}(\vartheta_2,\vartheta_1)) = 2) = P_{\vartheta_2}(X_1 = 1) = \vartheta_2,
\end{aligned}
$$

that

$$P_{\vartheta_1}(\varphi_\tau = 1) + P_{\vartheta_2}(\varphi_\tau = 0) = 2\,\vartheta_2 > 1,$$

i.e. (τ,φ) fails to be unbiased for $\{\vartheta_1\}$ against $\{\vartheta_2\}$. $\qquad\qquad\square$

Moreover, this simple example shows that even the additional assumption

$$\hat{t}_a = \hat{t}_{(g(a))}, \quad a \in A,$$

does not guarantee the unbiasedness of GSPPRT's.

But if $\hat{t}_a$ depends on a over the number of stages only, i.e. $\hat{t}_a = \hat{t}_{(h(a))}$, each GSPPRT can be shown to be unbiased. Unfortunately this condition, which obviously is fulfilled in the purely sequential case, is in general of minor practical importance.

(5.44) Theorem

Let (τ,φ) a GSPPRT with $\hat{t}_a = \hat{t}_{(h(a))}$ $\forall a \in A$ and

$$
\begin{aligned}
OC_n(\vartheta) &:= P_\vartheta(\varphi_\tau = 0, h(\tau) \le n) \\
PF_n(\vartheta) &:= P_\vartheta(\varphi_\tau = 1, h(\tau) \le n),
\end{aligned}
$$

$n \in \mathrm{I\!N}_0$. *Then*

$$OC_n(\vartheta_1) \ge OC_n(\vartheta_2), \quad PF_n(\vartheta_1) \le PF_n(\vartheta_2);$$

in particular, (τ,φ) is unbiased for $\{\vartheta_1\}$ against $\{\vartheta_2\}$.

The proof is essentially based on [Eg], p. 78-80 (a minor gap in that proof is filled according to a privat communication with K.-H. Eger): Simplifying the terminology by $q_a := q_a(\vartheta_2,\vartheta_1)$ we obtain for all $n \in \mathrm{I\!N}$

$$
\begin{aligned}
\{\varphi_\tau = 1, h(\tau) \le n\}^c &= \\
&= \{\varphi_\tau = 0, h(\tau) \le n-1\} + \{\varphi_{(\tau_1,\dots,\tau_n)} = 0, h(\tau) = n\} + \{h(\tau) > n\} \\
&= \{\varphi_\tau = 0, h(\tau) \le n-1\} + \{q_{(\tau_1,\dots,\tau_n)} < k_2^{(n)}, h(\tau) \ge n\}
\end{aligned}
$$

where the last step makes essentially use of $\hat{t}_a = \hat{t}_{(h(a))}$. This yields for all $\vartheta \in \Theta$ and $n \in \mathrm{I\!N}$

$$1 - PF_n(\vartheta) = OC_{n-1}(\vartheta) + P_\vartheta(q_{(\tau_1,\dots,\tau_n)} < k_2^{(n)}, h(\tau) \ge n),$$

moreover

$$PF_n(\vartheta) = PF_{n-1}(\vartheta) + P_\vartheta(q_{(\tau_1,\ldots,\tau_n)} \geq k_2^{(n)}, h(\tau) = n).$$

In strong analogy to the proof of Wald's inequalities (1.9)/(3.7) one obtains for all $n \in \mathbb{N}$

$$P_{\vartheta_2}(q_{(\tau_1,\ldots,\tau_n)} < k_2^{(n)}, h(\tau) \geq n) \ \leq \ k_2^{(n)} P_{\vartheta_1}(q_{(\tau_1,\ldots,\tau_n)} < k_2^{(n)}, h(\tau) \geq n)$$
$$P_{\vartheta_2}(q_{(\tau_1,\ldots,\tau_n)} \geq k_2^{(n)}, h(\tau) = n) \ \geq \ k_2^{(n)} P_{\vartheta_1}(q_{(\tau_1,\ldots,\tau_n)} \geq k_2^{(n)}, h(\tau) = n) :$$

The assertions are now shown by induction (over n): For $n = 0$ the (in)equalities are obviously fulfilled. For PF_n the step from $n - 1 \geq 0$ to n follows for $k_2^{(n)} \geq 1$ from

$$\begin{aligned} PF_n(\vartheta_2) \ &\geq \ PF_{n-1}(\vartheta_1) + k_2^{(n)} P_{\vartheta_1}(q_{(\tau_1,\ldots,\tau_n)} \geq k_2^{(n)}, h(\tau) = n) \\ &\qquad \text{by assumption and Wald's inequality} \\ &\geq \ PF_{n-1}(\vartheta_1) + P_{\vartheta_1}(q_{(\tau_1,\ldots,\tau_n)} \geq k_2^{(n)}, h(\tau) = n) \\ &= \ PF_n(\vartheta_1), \end{aligned}$$

for $k_2^{(n)} \leq 1$ from

$$\begin{aligned} 1 - PF_n(\vartheta_2) \ &\leq \ OC_{n-1}(\vartheta_1) + k_2^{(n)} P_{\vartheta_1}(q_{(\tau_1,\ldots,\tau_n)} < k_2^{(n)}, h(\tau) \geq n) \\ &\qquad \text{by assumption and Wald's inequality} \\ &\leq \ OC_{n-1}(\vartheta_1) + P_{\vartheta_1}(q_{(\tau_1,\ldots,\tau_n)} < k_2^{(n)}, h(\tau) \geq n) \\ &= \ 1 - PF_n(\vartheta_1); \end{aligned}$$

for OC_n one argues in an analogous way. $\qquad\qquad\square$

Although GSPPRT's may, in general, fail to be unbiased, one can give a sufficient condition for the monotonicity of the power function of GSPPRT's, which is fulfilled from a large class of examples. For this purpose we start with a condition which ensures the unbiasedness of these tests.

(5.45) Lemma

Let (τ, φ) be a GSPPRT and either $k_1^{(a)} \leq 1 \ \forall a \in A$ or $k_2^{(a)} \geq 1 \ \forall a \in A$. Then (τ, φ) is unbiased for $\{\vartheta_1\}$ against $\{\vartheta_2\}$. If either

$$k_1^{(a)} < 1 \qquad \forall a \in A \qquad and \qquad E_{\vartheta_1}\varphi_\tau \neq 1$$

or

$$k_2^{(a)} > 1 \qquad \forall a \in A \qquad and \qquad E_{\vartheta_2}\varphi_\tau \neq 0,$$

then (τ, φ) is strictly unbiased for $\{\vartheta_1\}$ against $\{\vartheta_2\}$.

Proof: Analogously to Wald's inequalities (1.9)/(3.7) follows

$$E_{\vartheta_1}\varphi_\tau = E_{\vartheta_2}((q_\tau(\vartheta_2,\vartheta_1))^{-1} \cdot \varphi_\tau) \leq E_{\vartheta_2}((k_2^{(\tau)})^{-1} \cdot \varphi_\tau)$$

and

$$E_{\vartheta_2}(1 - \varphi_\tau) \leq E_{\vartheta_1}(k_1^{(\tau)} \cdot (1 - \varphi_\tau)).$$

162

For $k_2^{(a)} \geq 1$ $\;\forall a \in A$ the first inequality yields $E_{\vartheta_1}\varphi_\tau \leq E_{\vartheta_2}\varphi_\tau$ and, under the additional assumption, $E_{\vartheta_1}\varphi_\tau < E_{\vartheta_2}\varphi_\tau$, i.e. unbiasedness and strict unbiasedness respectively. For the second case the assertions follow, by a corresponding argument, from the second inequality. $\qquad\qquad\qquad\qquad\qquad\qquad\qquad\qquad\qquad\qquad\qquad\qquad\qquad\qquad$ $\square$

(5.46) Theorem

Let $(P_\vartheta^{X_1})_{\vartheta \in \Theta}$ be a one-parameter exponential family in natural parametrization, i.e.

$$f_\vartheta(x) = h(x)\exp(\vartheta \cdot T(x) - \eta(\vartheta)),$$

and let (τ, φ) a GSPPRT with respect to ϑ_1, ϑ_2 such that there exists an $s \in (0, \infty)$ with $k_1^{(a)} \leq s^{g(a)} \leq k_2^{(a)}$ $\forall a \in A$. Then the power function PF of (τ, φ) is isotonic. If additionally

$$k_1^{(a)} < s^{g(a)} < k_2^{(a)} \qquad \forall a \in A$$

the PF is strictly isotonic on $\{\vartheta \in \Theta : 0 < E_\vartheta\varphi_\tau < 1\}$. In particular, for each SPPRT the power function is isotonic and strictly isotonic on $\{\vartheta \in \Theta : 0 < E_\vartheta\varphi_\tau < 1\}$.

Proof: Let $\vartheta' < \vartheta''$. It will be shown that (τ, φ) is a GSPPRT with respect to ϑ', ϑ'' which fulfills the conditions of (5.45): The likelihood ratio can be written in the form

$$
\begin{aligned}
q_a(\vartheta_2, \vartheta_1) &= \exp\!\left((\vartheta_2 - \vartheta_1)\sum_{i=1}^{g(a)} T(X_i) - (\eta(\vartheta_2) - \eta(\vartheta_1))g(a)\right) \\
&= H_{\vartheta_1,\vartheta_2}^{(g(a))}(T_{g(a)})
\end{aligned}
$$

where $H_{\vartheta_1,\vartheta_2}^{(n)}(x) := \exp((\vartheta_2 - \vartheta_1)x - (\eta(\vartheta_2) - \eta(\vartheta_1))n)$ and $T_n := \sum_{i=1}^{n} T(X_i)$. $H_{\vartheta_1,\vartheta_2}^{(n)}$ is, for each $n \in \mathbb{N}_0$, a strictly isotonic function; hence it follows

$$q_a(\vartheta_2, \vartheta_1) = H_{\vartheta_1,\vartheta_2}^{(g(a))} \circ (H_{\vartheta',\vartheta''}^{(g(a))})^{-1} \circ H_{\vartheta',\vartheta''}^{(g(a))} \circ T_{g(a)} = H^{(g(a))}(q_a(\vartheta'', \vartheta'))$$

where $H^{(n)} := H_{\vartheta_1,\vartheta_2}^{(n)} \circ (H_{\vartheta',\vartheta''}^{(n)})^{-1}$ is, for each $n \in \mathbb{N}_0$, a strictly isotonic function. Using the definitions

$$\tilde{t}_a := \hat{t}_a \circ H^{(g(a))}, \quad a \in A$$

one obtains

$$\tilde{t}_a(q_a(\vartheta'', \vartheta')) = \hat{t}_a(q_a(\vartheta_2, \vartheta_1)), \quad a \in A.$$

Therefore, (τ, φ) is a GSPPRT with respect to ϑ', ϑ''. To apply (5.45) further informations on the stopping bounds $\tilde{k}_1^{(a)}, \tilde{k}_2^{(a)}$ of $\tilde{t}_a$, $a \in A$, are needed. Obviously

$$\tilde{k}_i^{(a)} = (H^{(g(a))})^{-1}(k_i^{(a)}), \quad i = 1, 2.$$

Since

$$(H_{\vartheta_1,\vartheta_2}^{(n)})^{-1}(y) = \frac{\log y + (\eta(\vartheta_2) - \eta(\vartheta_1))n}{\vartheta_2 - \vartheta_1}$$

for all $n \in \mathbb{N}$, $y \in (0, \infty)$, it follows

$$(H^{(n)})^{-1}(y) = H^{(n)}_{\vartheta', \vartheta''} \circ (H^{(n)}_{\vartheta_1, \vartheta_2})^{-1}(y)$$
$$= \exp\left[\frac{\vartheta'' - \vartheta'}{\vartheta_2 - \vartheta_1} \log y + n \left(\frac{\vartheta'' - \vartheta'}{\vartheta_2 - \vartheta_1}(\eta(\vartheta_2) - \eta(\vartheta_1)) - (\eta(\vartheta'') - \eta(\vartheta'))\right)\right].$$

This yields the existence of a constant K, depending only on $s, \vartheta_1, \vartheta_2, \vartheta', \vartheta''$, such that

$$(H^{(n)})^{-1}(s^n) = \exp(n\, K).$$

If $K \geq 0$ then

$$\begin{aligned}
\tilde{k}_2^{(a)} &= (H^{(g(a))})^{-1}(k_2^{(a)}) \\
&\geq (H^{(g(a))})^{-1}(s^{g(a)}) \quad \text{according to the isotonicity of } H^{(n)} \\
&= \exp(g(a) \cdot K) \geq 1 \quad \forall a \in A.
\end{aligned}$$

For the case $K \leq 0$, one analogously concludes that $\tilde{k}_1^{(a)} \leq 1 \ \forall a \in A$. Under the additional assumptions one, moreover, obtains $\tilde{k}_2^{(a)} > 1 \ \forall a \in A$ or $\tilde{k}_1^{(a)} < 1 \ \forall a \in A$ respectively. Altogether, lemma (5.45) yields the assertion. $\qquad\square$

This result, and several other assertions of this chapter, indicate that the SPPRT's as treated in Chapter III are sequential decision procedures which are of practical interest as well as of theoretical importance.

Appendix A: Mathematical models for sequentially planned sampling procedures

In definition (1.22), we describe sampling procedures, where not only the total number of observations but also the sizes of sub-samples are determined sequentially, by the concept of sequential sampling plans. This description, which is essentially based on the concept of "control variables" by Haggstrom [Ha], is by no means the only possible one. In literature there exist several other concepts suited for our purposes; these have been developed nearly independent of each other. But it turns out that all these models are essentially "equivalent" (a detailed discussion was given by Duscha [Du]).

§ A.1 The concept of policies by Mandelbaum and Vanderbei

Mandelbaum and Vanderbei [M/V] considered a special case of the following situation[41]:

(A.1)

a) *Let $S \cup \{\hat{\infty}\}$ be a countable set, partially ordered by the relation $\hat{\preceq}$, having a minimal element $0 \in S$. Assume that*
   (i) $s \hat{\preceq} \hat{\infty} \quad \forall s \in S$
   (ii) S is locally finite, i.e. each $s \in S$ has only a finite number of predecessors. Denote by

$$D_s := \{q \in S : s \hat{\prec} q \text{ and } \not\exists\, r \in S : s \hat{\prec} r \hat{\prec} q\}$$

   the set of all direct successors of s.

b) *Let*
   - *$(\mathcal{X}, \mathcal{B}, P)$ be a probability space*
   - *$(\mathcal{F}_s)_{s \in S}$ be a filtration, i.e. a family $\mathcal{F}_s, s \in S$ of sub-σ-algebras of $\mathcal{B}$ such that $\mathcal{F}_r \subset \mathcal{F}_s$ for $r \hat{\preceq} s$.*
   - *$(Y_s)_{s \in S}$ be a family of integrable random variables adapted to $(\mathcal{F}_s)_{s \in S}$.*

$s \in S$ is interpreted as state of a system; from $s \in S$ one can reach, by corresponding decisions, direct successors $q \hat{\succ} s$. The set A_M of (1.20) endowed with the partial order $\preceq$ is an example of such a situation.

(A.2) Definition

a) *A stopping time (with respect to $(\mathcal{F}_s)_{s \in S}$) is a mapping $\nu : \mathcal{X} \to S \cup \{\hat{\infty}\}$ such that $\{\nu = s\} \in \mathcal{F}_s \quad \forall s \in S$.*

b) *For a stopping time ν*

$$\mathcal{F}_\nu := \{C \in \mathcal{B} : C \cap \{\nu = s\} \in \mathcal{F}_s \quad \forall s \in S\}$$

   is the σ-algebra of ν-history.

---

[41]Mandelbaum and Vanderbei additionally assume that the set D_s is finite for each $s \in S$.

c) A strategy with initial value $r \in S$ is a family $\sigma = (\sigma_n)_{n \in \mathbb{N}_0}$ of stopping times such that
(i) $r \equiv \sigma_0 \hat{\leq} \sigma_1 \hat{\leq} \ldots$
(ii) σ_{n+1} is $(\mathcal{F}_{\sigma_n}, \mathcal{P}(S))$- measurable for each $n \in \mathbb{N}_0$.

d) A policy with initial value $r \in S$ is a pair $\pi = (\sigma, N)$ where
(i) $\sigma = (\sigma_n)_{n \in \mathbb{N}_0}$ is a strategy with starting value r
(ii) N is a stopping time with respect to $(\mathcal{F}_{\sigma_n})_{n \in \mathbb{N}_0}$.

e) For a policy $\pi = ((\sigma_n)_{n \in \mathbb{N}_0}, N)$ define

$$\alpha(\pi) := \begin{cases} \sigma_N & on \ \{N < \infty\} \\ \hat{\infty} & elsewhere \end{cases}$$

and

$$Y_{\alpha(\pi)}(x) := \begin{cases} Y_s(x) & if \ \alpha(\pi)(x) = s \in S \\ -\infty & elsewhere; \end{cases}$$

for $r \in S$ let $\Pi_r := \{\pi : \pi \ policy \ with \ initial \ value \ r \ s.t. \ EY_{\alpha(\pi)} \ exists \ \}$.

Since S is countable

    - $\alpha(\pi)$ is a stopping time (with respect to $(\mathcal{F}_s)_{s \in S}$)

and

    - $Y_{\alpha(\pi)}$ is $(\mathcal{B}, \overline{\mathbb{B}})$- measurable.

Since $(\mathcal{F}_s)_{s \in S}$ is isotonic for each strategy $\sigma = (\sigma_n)_{n \in \mathbb{N}_0}$ one obtains

$$\mathcal{F}_{\sigma_n} \subset \mathcal{F}_{\sigma_{n+1}} \quad \forall n \in \mathbb{N}_0$$

(for comments on the procedure described by policies see [M/V], pp. 254/255).

One looks for an *optimal* policy i.e. a $\pi^* \in \Pi_0$ such that

$$E \, Y_{\alpha(\pi^*)} = \sup_{\pi \in \Pi_0} E \, Y_{\alpha(\pi)}.$$

Since $Y_{\alpha(\pi)} = -\infty$ on $\{\alpha(\pi) = \hat{\infty}\}$, respectively $\{N = \infty\}$, one may restrict attention to the case $P(\{N < \infty\}) = 1$. For further reductions of Π_0 we use the following notation:

(A.3) Definition
A policy $\pi = ((\sigma_n)_{n \in \mathbb{N}_0}, N)$ is called successive, iff
(i) $N = \inf\{n \in \mathbb{N}_0 : \sigma_{n+1} = \sigma_n\}$ (where $\inf \emptyset := \infty$)
(ii) $\sigma_{n+1}(x) \in D_{\sigma_n(x)} \ \forall n \in \mathbb{N}_0, x \in \{n < N\}$
(iii) $\sigma_{n+1} = \sigma_n \quad \forall n \in \mathbb{N}_0 \ on \ \{n \geq N\}$

Connecting arbitrary $r, q \in S, r \hat{\leq} q$, by finite paths $r = r_0, r_1, \ldots, r_j = q$ of direct successors one obtains, that for each policy $\pi \in \Pi_s, s \in S$, there exists a successive policy $\tilde{\pi} \in \Pi_s$ such that $\alpha(\tilde{\pi}) \equiv \alpha(\pi)$. Hence one may, for each initial value $s \in S$, restrict attention to

$$\tilde{\Pi}_s := \{\pi \in \Pi_s : \pi \ successive, \ E \, Y^-_{\alpha(\pi)} < \infty\}.$$

166

Our aim is to prove relations between the concepts of policies and of sampling plans. Let for each $s \in S$ s.t. $D_s \neq \emptyset$

$$\delta_s : \{1, \ldots, |D_s|\} \to D_s$$

be a bijective mapping, and define inductively

$$\varphi(()) \; := \; 0 \; (\in S), \tilde{M}_{()} := \begin{cases} \{1, \ldots, |D_0|\} & \text{if } D_0 \neq \emptyset \\ \emptyset & \text{elsewhere} \end{cases} \;,$$

$$\varphi(a) \; := \; \delta_{\varphi((a_1,\ldots,a_{j-1}))}(a_j),$$

$$\tilde{M}_a \; := \; \begin{cases} \{1, \ldots, |D_{\varphi(a)}|\} & \text{if } D_{\varphi(a)} \neq \emptyset \\ \emptyset & \text{elsewhere} \end{cases}$$

for $a = (a_1, \ldots, a_j) \in \tilde{M}_{()} \times \tilde{M}_{(a_1)} \times \ldots \times \tilde{M}_{(a_1,\ldots,a_{j-1})}$; furthermore

$$\tilde{\mathcal{A}} := \bigcup_{j=1}^{\infty} \{a = (a_1, \ldots, a_j) : a_i \in \tilde{M}_{a^{i-1}}, 1 \leq i \leq j\} \cup \{()\}.$$

Obviously the mapping $\varphi : \tilde{\mathcal{A}} \to S$ is surjective (but not necessarily injective; if S is a tree, φ is also injective).

Now consider the problem of optimal sampling

$$((\mathcal{X}, \mathcal{B}, P), \; \mathcal{A}, (\mathcal{B}_a)_{a \in A}, \; (Z_a)_{a \in A})$$

defined by
- $(\mathcal{X}, \mathcal{B}, P)$ as in (A.1)b)
- $\mathcal{A} := \tilde{\mathcal{A}}$, $A := A_M$ where $M := \text{IN}$
- $\mathcal{B}_a := \mathcal{F}_{\varphi(a)}$ $\forall a \in \mathcal{A}$ and $\mathcal{B}_a := \mathcal{B}$ $\forall a \in A \backslash \mathcal{A}$
- $Z_a := Y_{\varphi(a)}$ $\forall a \in \mathcal{A}$.

(A.4) Lemma

> *Consider $s \in S$ and $a \in \mathcal{A}$ s.t. $\varphi(a) = s$; then for each policy $\pi \in \tilde{\Pi}_s$ there exists a sampling plan $\tau \in T_a$ such that*
>
> $$Z_\tau(x) = Y_{\alpha(\pi)}(x) \qquad \forall x \in \mathcal{X}.$$

Proof: Let $a = (a_1, \ldots, a_j) \in \mathcal{A}, s \in S, \varphi(a) = s$, and $\pi = ((\sigma_n)_{n \in \text{IN}_0}, N) \in \tilde{\Pi}_s$. Then

$$\tau(x) := \begin{cases} (a_1, \ldots, a_j) & \text{if } N(x) = 0 \\ (a_1, \ldots, a_j, \delta_s^{-1}(\sigma_1(x)), \delta_{\sigma_1(x)}^{-1}(\sigma_2(x)), \ldots, \delta_{\sigma_{N(x)-1}(x)}^{-1}(\sigma_{N(x)}(x))) \\ \qquad \text{if } 0 < N(x) < \infty \\ (a_1, \ldots, a_j, \delta_s^{-1}(\sigma_1(x)), \delta_{\sigma_1(x)}^{-1}(\sigma_2(x)), \ldots) & \text{if } N(x) = \infty \end{cases}$$

defines a sampling plan $\tau : \mathcal{X} \to A \cup A^*$ (cf. (1.22)). Obviously $\tau \succeq a$ and $\{\tau \in A\} = \{N < \infty\}$; hence $P(\{\tau \in A\}) = 1$ and

$$E \, Z_\tau = E \, Y_{\varphi \circ \tau} = E \, Y_{\alpha(\pi)} > -\infty, \; i.e. \; \tau \in T_a.$$

Moreover

$$Z_\tau = Y_{\varphi \circ \tau} = Y_{\alpha(\pi)} \quad \text{on} \quad \{\tau \in A\} = \{N < \infty\}$$

and

$$Z_\tau = -\infty = Y_{\alpha(\pi)} \quad \text{on} \quad \{\tau \notin A\} = \{N = \infty\}. \qquad \square$$

Conversely we obtain:

(A.5) Lemma

Consider $a \in \mathcal{A}$ and $s \in S$ s.t. $\varphi(a) = s$; then for each sampling plan $\tau \in T_a$ there exists a policy $\pi \in \tilde{\Pi}_s$ such that

$$Y_{\alpha(\pi)} = Z_\tau \qquad P - a.s.$$

Proof: Let $a = (a_1, \ldots, a_j) \in \mathcal{A}, s \in S, \varphi(a) = s$, and $\tau \in T_a$. Then

$$\hat{\tau}(x) := \begin{cases} \tau(x) & \text{if } \tau(x) \in \mathcal{A} \text{ or } \tau(x) \in \mathcal{A}^\star := \{c \in A^\star : c_{i+1} \in \tilde{M}_{c^i} \ \forall i \in \mathbb{N}_0\} \\ b^i & \text{if } \tau(x) = b \notin \mathcal{A} \cup \mathcal{A}^\star \text{ and} \\ & i = \inf\{0 \le \ell \le h(b) : \tau(x) \succeq b^{\ell+1}, b^{\ell+1} \notin \mathcal{A}\} < \infty \end{cases}$$

(where $h(b) := \infty$ for $b \in A^\star$) defines a sampling plan $\hat{\tau}$ such that

$$P(\{\hat{\tau} \in \mathcal{A}, \hat{\tau} = \tau\}) \ge P(\{\tau \in \mathcal{A}\}) = 1,$$

hence $Z_{\hat{\tau}} = Z_\tau$ P-a.s. Defining

$$N(x) \quad := \quad h(\hat{\tau}(x)) - j \quad \text{for } x \in \mathcal{X}$$

$$\sigma_0 \quad :\equiv \quad s$$

$$\sigma_n(x) \quad := \quad \begin{cases} \varphi((\hat{\tau}(x))^{j+n}) & \text{if } n < h(\hat{\tau}(x)) - j \\ \\ \varphi(\hat{\tau}(x)) & \text{elsewhere} \end{cases} \qquad \text{for} \quad n \in \mathbb{N}, x \in \mathcal{X}$$

yields a successive policy $\pi := ((\sigma_n)_{n \in \mathbb{N}_0}, N)$ with starting value s. Obviously

$$\{N < \infty\} = \{\hat{\tau} \in \mathcal{A}\},$$
$$Z_{\hat{\tau}}(x) = Y_{\varphi(\hat{\tau}(x))}(x) = Y_{\sigma_{N(x)}(x)}(x) = Y_{\alpha(\pi)}(x) \text{ on } \{\tau \in \mathcal{A}\},$$

and

$$Z_{\hat{\tau}}(x) = -\infty = Y_{\alpha(\pi)}(x) \text{ on } \{\hat{\tau} \notin \mathcal{A}\} = \{N = \infty\}.$$

Altogether we obtain

$$Y_{\alpha(\pi)} = Z_{\hat{\tau}} = Z_\tau \quad P - a.s.;$$

in particular $E\, Y_{\alpha(\pi)} = E\, Z_\tau > -\infty$, i.e. $\pi \in \tilde{\Pi}_s$. $\qquad \square$

(A.4) and (A.5) show the equivalence of both concepts:

(A.6) Corollary

Let $s \in S$ and $a \in \mathcal{A}$ such that $\varphi(a) = s$; then

$$\sup_{\pi \in \tilde{\Pi}_s} E\, Y_{\alpha(\pi)} = \sup_{\tau \in T_a} E\, Z_\tau;$$

in particular $\sup_{\pi \in \Pi_0} E\, Y_{\alpha(\pi)} = \sup_{\tau \in T} E\, Z_\tau$.

As a consequence, the results of chapter II can be used to solve the corresponding problems for policies: Analogously to p. 35 we assume that

$$E(\sup_{s \in S} Y_s^+) < \infty.$$

According to (A.6), for $s \in S$ and $a \in \mathcal{A}$ such that $\varphi(a) = s$ the essential suprema

$$V_s := \operatorname*{ess\,sup}_{\pi \in \Pi_s} E(Y_{\alpha(\pi)}|\mathcal{F}_s)$$

and

$$U_a(:= \operatorname*{ess\,sup}_{\tau \in T_a} E(Z_\tau|\mathcal{B}_a))$$

coincide P-a.s.. Hence lemma (2.8) yields a generalization of theorem 1 in [M/V]:

(A.7) Lemma *(Bellman-Equation)*
  For each $s \in S$
$$V_s = \max\{Y_s, \sup_{q \in D_s} E(V_q|\mathcal{F}_s)\} \quad P - a.s.$$

Analogously to (2.9) the values $V_s, s \in S$, can be used to construct a candidate for an optimal policy: For $s \in S$ define

$$C_s := \{E(V_r|\mathcal{F}_s) < \sup_{q \in D_s} E(V_q|\mathcal{F}_s) \quad \forall r \in D_s\}$$

(where $C_s = \emptyset$ for $s \in S$ s.t. $D_s = \emptyset$), and for $r \in D_s$

$$L(s,r) := \{\delta_s^{-1}(r) = \inf\{j \in \{1,\ldots,|D_s|\} : E(V_{\delta_s(j)}|\mathcal{F}_s) = \sup_{q \in D_s} E(V_q|\mathcal{F}_s)\}\} \cap \{V_s > Y_s\};$$

moreover, let $\sigma_0^\star :\equiv 0$ and for $n \in \mathbb{N}_0$

$$\sigma_{n+1}^\star(x) := \begin{cases} s & \text{if } x \in L(\sigma_i^\star(x), \sigma_{i+1}^\star(x)) \text{ for } \forall 0 \le i < n \text{ and } x \in L(\sigma_n^\star, s) \\ \sigma_n^\star(x) & \text{if } \exists i : 0 \le i \le n \text{ s.t. } x \in L(\sigma_j^\star(x), \sigma_{j+1}^\star(x)) \\ & \qquad \forall 0 \le j < i, V_{\sigma_i^\star(x)}(x) = Y_{\sigma_i^\star(x)}(x) \\ t_{n+1-i} & \text{if } \exists i : 0 \le i \le n \text{ s.t. } x \in L(\sigma_j^\star(x), \sigma_{j+1}^\star(x)) \; \forall 0 \le j < i, \\ & \qquad V_{\sigma_i^\star(x)}(x) > Y_{\sigma_i^\star(x)}(x), x \in C_{\sigma_i^\star(x)}, \\ & \qquad \text{where } t_0 := \sigma_i^\star(x), \; t_{k+1} := \delta_{t_k}(1) \; \forall k \in \mathbb{N}_0, \end{cases}$$

$$N^\star := \inf\{n \in \mathbb{N}_0 : \sigma_{n+1}^\star = \sigma_n^\star\} \text{ (where inf } \emptyset := \infty).$$

Then $\pi^\star := ((\sigma_n^\star)_{n \in \mathbb{N}_0}, N^\star)$ is a successive policy with starting value 0. Applying $\pi^\star$ we start with $\sigma_0^\star \equiv 0$ and stop as soon as $\sigma_n^\star(x)$ equals its successor $\sigma_{n+1}^\star(x)$, i.e. the actual pay-off $Y_{\sigma_n^\star(x)}(x)$ coincides with the maximal (conditionally) expected pay-off $V_{\sigma_n^\star(x)}(x)$. For $V_{\sigma_n^\star(x)}(x) > Y_{\sigma_n^\star(x)}(x)$ the successor $\sigma_{n+1}^\star(x)$ is determined by the condition $x \in L(\sigma_n^\star(x), \sigma_{n+1}^\star(x))$, provided that the supremum in the Bellman-equation (A.7) is attained; otherwise (i.e. for $x \in C_{\sigma_n^\star(x)}$) for each further step the respective direct successor with the smallest "index" is chosen. This policy $\pi^\star$ fulfills

$$\sigma_i^\star = \varphi(\tau^{\star i}) \quad P - a.s. \quad \forall i \in \mathbb{N}_0$$

and

$$N^* = h \circ \tau^* \quad \text{on} \quad \{N^* < \infty\} = \{\tau^* \in \mathcal{A}\}$$

where τ^* denotes the sampling plan defined in (2.9). Therefore one obtains

$$\alpha(\pi^*) = \varphi(\tau^*) \text{ on } \{N^* < \infty\} \text{ and } P(N^* < \infty) = P(\tau^* \in A),$$

hence $E\, Y_{\alpha(\pi^*)} = E\, Z_{\tau^*}$.

Corollary (A.6) together with theorem (2.10) now yields a generalization of theorem 2 of Mandelbaum/Vanderbei [M/V]:

(A.8) Theorem

If $N^* < \infty$ P-a.s. (i.e. $\pi^* \in \tilde{\Pi}_0$), then π^* is optimal (in Π_0).

In a similar way assertion (2.14) can be utilized to obtain an analogous sufficient condition for the optimality of π^*:

(A.9) Theorem

Let $P(\{V_s > Y_s\} \cap C_s) = 0 \quad \forall s \in S$, let X be an integrable random variable and $(r_n)_{n \in \mathbb{N}_0}$ be a sequence of real numbers such that $\lim_{n \to \infty} r_n = -\infty$ and

$$Y_s \leq X + r_n \quad P - a.s.$$

for all $s \in S$ for which the longest path of direct successors between 0 and s has length n. Then π^* is an optimal policy (in Π_0).

Corresponding optimality assertions can be obtained for the case that an arbitrary direct successor is chosen among those $r \in D_s$ for which

$$E(V_r|\mathcal{F}_s) = \sup_{q \in D_s} E(V_q|\mathcal{F}_s)$$

(provided the supremum is achieved). Hence our results hold, in general, for a whole class of strategies.

For sampling plans the optimality results may considerably be simplified if the pay-off process $(Z_s)_{s \in S}$ has a "Markov structure" (see Ch. II, § 1). Corresponding assertions hold true for (our generalizations of) the results of Mandelbaum and Vanderbei [M/V].

(A.10) Definition

    a) A stochastic process $(X_s)_{s \in S}$ (with values in $(\mathcal{Y},\mathcal{C})$) is a Markov process (with respect to $(\mathcal{F}_s)_{s \in S}$) if

$$P(X_r \in C|\mathcal{F}_s) = P(X_r \in C|X_s)\ P - a.s.\ \forall s \in S, r \in D_s, C \in \mathcal{C}.$$

    b) A Markov process $(X_s)_{s \in S}$ is stationary if $|D_s| = m \in \overline{\mathbb{N}}\ \forall s \in S$ and if for each $k \in \{1, \ldots, m\}$ there exists a transition kernel Q_k on $(\mathcal{Y},\mathcal{C})$ such that

$$Q_k(X_s, C) = P(X_{\delta_s(k)} \in C|\mathcal{F}_s)\ P - a.s.\ \forall s \in S, C \in \mathcal{C}.$$

    If $X_0 = y$ P-a.s., y is called initial state of $(X_s)_{s \in S}$.

c) For the general situation (A.1) we have the Markov case if

$$Y_s = \eta_s \circ X_s \quad \forall s \in S$$

where $(X_s)_{s \in S}$ is a Markov process (with respect to $(\mathcal{F}_s)_{s \in S}$) and $\eta_s, s \in S$, are measurable functions.

In the Markov case we obtain

$$P(Y_r \in C | \mathcal{F}_s) = P(Y_r \in C | X_s) \ P - a.s. \ \forall s \in S, r \in D_s, C \in \overline{\mathbb{B}^1}$$

i.e. the stochastic behaviour after "time" s depends on X_s only. For the corresponding problem of optimal sampling the sequence $(X_{\varphi(a)})_{a \in \mathcal{A}}$ is a Markov process with respect to $(\mathcal{B}_a)_{a \in \mathcal{A}}$; since moreover

$$Z_a = Y_{\varphi(a)} = \eta_{\varphi(a)} \circ X_{\varphi(a)} \ \forall a \in \mathcal{A}$$

we have the Markov case (see (2.17) b)). Therefore theorem (2.23) can be applied:

(A.11) Remark

In the Markov case with finite S there exist $(\sigma(X_s), \overline{\mathbb{B}^1})$-measurable versions of V_s and $E(V_r | \mathcal{F}_s) \ \forall s \in S, r \in D_s$.

This may partially be extended to the case of arbitrary S: If for

$$V_s^n := \operatorname*{ess\,sup}_{\pi \in \Pi_s^n} E(Y_{\alpha(\pi)} | \mathcal{F}_s), \ s \in S, n \in \mathbb{N}_0$$

where

$$\tilde{\Pi}_s^n := \{\pi = ((\sigma_j)_{j \in \mathbb{N}_0}, N) \in \tilde{\Pi}_s : \sum_{i=1}^{N(x)} \delta_{\sigma_{i-1}(x)}^{-1}(\sigma_i(x)) \leq n \ \forall x \in \mathcal{X}\}$$

$\lim_{n \to \infty} V_s^n = V_s \ P - a.s. \ \forall s \in S$ then the monotone convergence theorem yields the desired result (see (2.25)).

Further simplifications are possible if the underlying Markov process $(X_s)_{s \in S}$ is stationary and the index set S has an additional structure (which is obviously fulfilled for trees).

(A.12) Definition

(S, φ) has Markov structure (for the definition of φ see p. 220) if
(i) $|D_s| = m \in \overline{\mathbb{N}} \quad \forall s \in S$
(ii) $\varphi(a) = \varphi(\hat{a}) \Leftrightarrow \varphi(ba) = \varphi(\hat{b}\hat{a}) \forall a, \hat{a}, b, \hat{b} \in \mathcal{A} : \varphi(b) = \varphi(\hat{b})$.

If (S, φ) has Markov structure then for any $r, s \in S$ s.t. $s = \varphi(a)$, $r = \varphi(b)$ there exists a uniquely determined $q \overset{.}{\succeq} r$ such that $q = \varphi(ba)$. Hence we may define a mapping $\gamma : S^2 \to S$ by

$$\gamma(r, s) := \varphi(ba).$$

Obviously $\gamma(0, s) = s = \gamma(s, 0) \ \forall s \in S$ and, for $r, s, q \in S, \gamma(r, q) \in D_{\gamma(r,s)}$ iff $q \in D_s$. Hence we are able to "merge" any two states $r, s \in S$ in an analogous way as it is done in the concept of sampling plans by means of $(b, a) \mapsto ba$.

In the sequel we assume that (S, φ) has Markov structure, that $(X_s^{(y)})_{s \in S}$ is a stationary Markov process with initial state $y \in \mathcal{Y}$ and transition kernels $Q_k, k \in \{1, \ldots, m\}$ independent of $y \in \mathcal{Y}$, and that $Y_s^{(y)} = \eta_s \circ X_s^{(y)}$ is the pay-off function (i.e. for each initial state y we have the Markov case). As usual in the context of optimal stopping we assume that

$$E(\eta_s^r \circ X_s^{(y)}) > -\infty \ \forall s \in S \ \text{ and } \ E(\sup_{s \in S}(\eta_s^r \circ X_s^{(y)})^+) < \infty$$

for all $r \in S, y \in \mathcal{Y}$, where $\eta_s^r := \eta_{\gamma(r,s)}$. Hence the definitions

$$V_s^{y,n} := \operatorname*{ess\,sup}_{\pi \in \Pi_s^n} E(Y_{\alpha(\pi)}^{(y)} | \mathcal{F}_s)$$

and

$$w_{s,j,n}(y) := \sup_{\pi \in \tilde{\Pi}_{\varphi((j))}^n} E(\eta_{\alpha(\pi)}^s \circ X_{\alpha(\pi)}^{(y)}), j \in M,$$

make sense (for all $s \in S, y \in \mathcal{Y}$ and $n \in \mathbb{N}_0$; $\eta_{\alpha(\pi)}^s \circ X_{\alpha(\pi)}^{(y)} := -\infty$ on $\{\alpha(\pi) = \hat{\infty}\}$ for each policy π).

Then the structure of optimal policies may be simplified according to

(A.13) Theorem
$$E(V_r^{y,n} | \mathcal{F}_s) = w_{s,\delta_s^{-1}(r),n}(X_s^{(y)}) \quad P - a.s.$$
$$\text{for all } n \in \mathbb{N}_0, \ s \in S, \ r \in D_s \text{ and } y \in \mathcal{Y}.$$

Proof: For fixed $y \in \mathcal{Y}$ consider the corresponding problem of optimal sampling with pay-off $Y_{\varphi(a)}^{(y)} = \eta_{\varphi(a)} \circ X_{\varphi(a)}^{(y)}$. Since $(X_{\varphi(a)}^{(y)})_{a \in A}$ is a stationary Markov process (with initial state y) and (according to (A.6))

$$V_r^{y,n} = U_{aj}^{y,n+g(aj)} \quad P - a.s.$$

$\forall n \in \mathbb{N}_0, r, s \in S$ and $a \in A, j \in M$ s.t. $\varphi(a) = s, \varphi(aj) = r$, theorem (2.26) yields

$$E(V_r^{y,n} | \mathcal{F}_s)(x) = E(U_{aj}^{y,n+g(aj)} | \mathcal{F}_{\varphi(a)})(x)$$
$$= v_{a,j,n+j}(X_{\varphi(a)}^{(y)}(x)) = \sup_{\tau \in T_{(j)}^{n+j}} E(\eta_{\varphi(a\tau)} \circ X_{\varphi(\tau)}^{(z)}) \text{ where } z := X_{\varphi(a)}^{(y)}(x)$$
$$= \sup_{\tau \in T_{(j)}^{n+j}} E(\eta_{\varphi(\tau)}^s \circ X_{\varphi(\tau)}^{(z)})$$
$$\text{since } \varphi(ab) = \gamma(s, \varphi(b)) \text{ and therefore } \eta_{\varphi(ab)} = \eta_{\varphi(h)}^s \ \forall b \in A$$
$$= \sup_{\pi \in \tilde{\Pi}_{\varphi((j))}^n} E(\eta_{\alpha(\pi)}^s \circ X_{\alpha(\pi)}^{(z)}) \text{ according to (A.6)}$$
$$= w_{s,j,n}(z) = w_{s,\delta_s^{-1}(r),n}(X_s^{(y)}(x)) \quad P - a.s. \qquad \square$$

This assertion means that under our "Markovian" assumptions optimal strategies may be determined which depend only on the present state of the system.

Applying the monotone convergence theorem for conditional expectations yields analogous results (corresponding to (2.27)) for the "infinite horizon" case.

Moreover, if also the functions η_s are "stationary", i.e. independent of s, the structure of optimal strategies may even be simplified to a "first entrance/j-continuation set"-type (analogously to (2.28); for details see [Du]).

§ A.2 The concept of tactics by Krengel and Sucheston

Starting point for Krengel and Sucheston [K/S] was the same situation as in (A.1). Their concept of an "optimal choice of successors" is given by the following notion:

(A.14) Definition

a) *For $s \in S$ let*
$$S_s := \{r \in S : s \overset{\curvearrowright}{\preceq} r\} \quad and \quad \overline{S_s} := S_s \cup \{\hat{\infty}\},$$
and let $\sum_s$ be the set of all stopping rules $\nu : \mathcal{X} \to \overline{S_s}$.

b) *A tactic with starting point $s \in S$ is a family*
$$\mathcal{H}_s := \{H_{r,q} : r \in S_s, q \in D_r \cup \{r\}\}$$
such that
(i) $H_{r,q} \in \mathcal{F}_r \; \forall r \in S_s, q \in D_r \cup \{r\}$,
(ii) $\sum_{q \in D_r \cup \{r\}} H_{r,q} = \mathcal{X} \; \forall r \in S_s$.

For each $s \in S$ and $r \overset{\curvearrowright}{\succeq} s$ the sets $H_{r,q}$ of a tactic $\mathcal{H}_s$ form a partition of $\mathcal{X}$; for $x \in H_{r,q}$ one decides to move from the point $r \in S_s$ to the direct "successor" $q \in D_r \cup \{r\}$.

A tactic $\mathcal{H}_s, s \in S$, generates a stopping rule $\tau_{\mathcal{H}_s} \in \sum_s$ as follows

(i) If $x \in \bigcap_{i=0}^n H_{r_i,r_{i+1}}$ where $n \in \mathbb{N}_0, H_{r_i,r_{i+1}} \in \mathcal{H}_s, 0 \le i \le n$, $r_{i+1} \in D_{r_i}, 0 \le i \le n-1, r_0 = s, r_{n+1} = r_n$, then $\tau_{\mathcal{H}_s}(x) := r_n$.

(ii) If $x \in \bigcap_{i=0}^\infty H_{r_i,r_{i+1}}$ where $H_{r_i,r_{i+1}} \in \mathcal{H}_s, r_{i+1} \in D_{r_i} \; \forall i \in \mathbb{N}_0$ and $r_0 = s$ then $\tau_{\mathcal{H}_s}(x) := \hat{\infty}$.

For a tactic $\mathcal{H}_s$ (with associated stopping rule $\tau_{\mathcal{H}_s}$) let
$$Y_{\mathcal{H}_s} := \begin{cases} Y_r & \text{on } \{\tau_{\mathcal{H}_s} = r\}, r \in S_s \\ -\infty & \text{on } \{\tau_{\mathcal{H}_s} = \hat{\infty}\} \end{cases}$$

be the pay-off at the "end" of the procedure, and let for $s \in S$
$$K_s := \{\mathcal{H}_s : \mathcal{H}_s \text{ tactic with starting point } s \text{ s.t. } E\,Y_{\mathcal{H}_s} \text{ exists}\}$$

denote the set of all tactics for which the expected pay-off is defined. Again we look for an *optimal* procedure i.e. a tactic $\mathcal{H}_0^* \in K_0$ such that
$$E\,Y_{\mathcal{H}_0^*} = \sup_{\mathcal{H}_0 \in K_0} E\,Y_{\mathcal{H}_0}.$$

It turns out that this aim can be achieved utilizing strong relationships between tactics and policies.

(A.15) Lemma
For each tactic $\mathcal{H}_s \in K_s, s \in S$, there exists a policy $\pi \in \Pi_s$ such that
$$Y_{\alpha(\pi)}(x) = Y_{\mathcal{H}_s}(x) \quad \forall x \in \mathcal{X}.$$

Proof: Let $\mathcal{H}_s = \{H_{r,q} : r \in S_s, q \in D_r \cup \{r\}\} \in \mathcal{K}_s$. Defining inductively

$$\sigma_0 :\equiv r, \ \sigma_n(x) := r \text{ for } n \in \mathbb{N}, \ x \in H_{\sigma_{n-1}(x),r}$$

$$N := \inf\{n \in \mathbb{N}_0 : \sigma_n = \sigma_{n+1}\}$$

yields a policy $\pi = ((\sigma_n)_{n \in \mathbb{N}_0}, N)$ with initial value s. By construction $\alpha(\pi)(x) = \tau_{\mathcal{H}_s}(x)$ $\forall x \in \mathcal{X}$, hence $Y_{\alpha(\pi)}(x) = Y_{\mathcal{H}_s}(x) \ \forall x \in \mathcal{X}$. In particular we obtain $E\,Y_{\alpha(\pi)} = E\,Y_{\mathcal{H}_s}$, i.e. $\pi \in \Pi_s$. $\qquad\square$

Conversely we obtain

(A.16) Lemma

For each policy $\pi \in \Pi_s, s \in S$, there exists a tactic $\mathcal{H}_s \in \mathcal{K}_s$ such that

$$Y_{\mathcal{H}_s}(x) = Y_{\alpha(\pi)}(x) \quad \forall x \in \mathcal{X}.$$

Proof: According to section A.1 one may restrict attention to successive policies. For such a policy $\pi = ((\sigma_n)_{n \in \mathbb{N}_0}, N) \in \Pi_s$ define

$$H_{r,q} := \sum_{n \in \mathbb{N}_0} \{\sigma_n = r, \sigma_{n+1} = q\}, r \in S_s, q \in D_r$$

and

$$H_{r,r} := \bigcup_{n \in \mathbb{N}_0} \{\sigma_n = \sigma_{n+1} = r\} + (\bigcup_{n \in \mathbb{N}_0} \{\sigma_n = r\})^c, r \in S_s.$$

Then $\mathcal{H}_s := \{H_{r,q} : r \in S_s, q \in D_r \cup \{r\}\}$ is a tactic with starting point s and

$$\{\tau_{\mathcal{H}_s} = r\} = \bigcup_{n \in \mathbb{N}_0} \sum_{\substack{s=r_0,r_1,\ldots,r_n=r_{n+1}=r \\ r_i \in D_{r_i}, 0 \le i < n}} \bigcap_{i=0}^{n} H_{r_i,r_{i+1}}$$

$$= \bigcup_{n \in \mathbb{N}_0} \{N = n, \sigma_n = r\} = \{\alpha(\pi) = r\} \quad \forall r \in S_s$$

as well as $\{\tau_{\mathcal{H}_s} = \hat{\infty}\} = \{\alpha(\pi) = \hat{\infty}\}$. Therefore $Y_{\mathcal{H}_s}(x) = Y_{\alpha(\pi)}(x) \ \forall x \in \mathcal{X}$; in particular we obtain $E\,Y_{\mathcal{H}_s} = E\,Y_{\alpha(\pi)}$, i.e. $\mathcal{H}_s \in \mathcal{K}_s$. $\qquad\square$

These assertions show the equivalence of both concepts, which yields as a special consequence:

(A.17) Corollary

$$\sup_{\mathcal{H}_s \in \mathcal{K}_s} E\,Y_{\mathcal{H}_s} = \sup_{\pi \in \Pi_s} E\,Y_{\alpha(\pi)} \quad \forall s \in S;$$

in particular

$$\sup_{\mathcal{H}_0 \in \mathcal{K}_0} E\,Y_{\mathcal{H}_0} = \sup_{\pi \in \Pi_0} E\,Y_{\alpha(\pi)}.$$

Therefore we can apply the optimality assertions for policies to obtain corresponding results for tactics: Under the general assumption

$$E(\sup_{s \in S} Y_s^+) < \infty$$

(A.17) immediately yields

$$V_s = \operatorname*{ess\,sup}_{\mathcal{H}_s \in \mathcal{K}_s} E(Y_{\mathcal{H}_s}|\mathcal{F}_s) =: U_s \quad \forall s \in S$$

(for the definition of V_s see p. 223); lemma (A.7) leads, therefore, to the following generalization of lemma 1.1 of Krengel/Sucheston [K/S][42].

(A.18) Lemma *(Bellman-Equation)*

$$U_s = \max\{Y_s, \sup_{r \in D_s} E(U_r|\mathcal{F}_s)\} \quad P - a.s. \ \forall s \in S.$$

Now it is evident how to obtain an *optimal tactic*: If the policy $\pi^\star$ given on p. 168 is optimal (see theorem (A.8)), construct a tactic $\mathcal{H}_0^\star \in \mathcal{K}_0$ according to (A.16), i.e. substitute the values V_s in $\pi^\star$ for $U_s, s \in S$, in order to determine the sets $H_{r,q}^\star, r \in S, q \in D_r \cup \{r\}$. Then $\pi^\star$ and $\mathcal{H}_0^\star$ have the same pay-offs, and the optimum property of $\pi^\star$ carries over to $\mathcal{H}_0^\star$.

Moreover, utilizing the structural results for the Markov case we obtain, under appropriate "Markovian" assumptions, analogous results for tactics.

§ A.3 The concept of decision functions by Washburn and Willsky

The situation in the concept of decision functions by Washburn and Willsky [Wa/Wi] is much the same as that considered in the previous ones[43]; again we make the assumptions described in (A.1).

(A.19) Definition
   A decision function with starting point $s \in S$ is a function $\Psi_s : S_s \times \mathcal{X} \to S_s$
   (for the definition of S_s see (A.14)) such that
   (i) $r \stackrel{\hat{}}{\preceq} \Psi_s(r, x) \quad \forall r \in S_s, x \in \mathcal{X}$
   (ii) $\Psi_s(r, \cdot)$ is $(\mathcal{F}_r, \mathcal{P}(S))$-measurable $\forall r \in S_s$.

Here $\Psi_s(r, x)$ denotes the successor of r to be chosen next; the measurability condition means that this choice may depend on the available information only.

For a decision function Ψ_s with starting point s let $\Psi_s^0(x) := s \quad \forall x \in \mathcal{X}$ and

$$\Psi_s^{n+1}(s) := \Psi_s(\Psi_s^n(x), x) \quad \forall x \in \mathcal{X}, \ n \in \mathbb{N}_0.$$

Each of these mappings $\Psi_s^n, n \in \mathbb{N}_0$, is a stopping rule with respect to $(\mathcal{F}_r)_{r \in S}$. As soon as $\Psi_s^n(x) = \Psi_s^{n+1}(x)$ for an $n \in \mathbb{N}_0$, the sequence becomes stationary, i.e. $\Psi_s^k(x) = \Psi_s^n(x) \quad \forall k \geq n$. This leads to

---

[42]Krengel and Sucheston consider only the case of finite S.

[43]For the general definitions they do without the existence of a root 0 in S and of an element $\hat{\infty}$ and they make no assumptions on the number of predecessors or direct successors of elements $s \in S$; but for their optimality results they also assume additional properties (uniform boundedness of the pay-off, finitely many immediate successors etc.).

(A.20) Definition

For a decision function Ψ_s

$$\tau_{\Psi_s} := \inf\{n \in \mathbb{N}_0 : \Psi_s^n = \Psi_s^{n+1}\}$$

is called the corresponding stopping rule.

Since

$$\{\tau_{\Psi_s} = n\} \cap \{\Psi_s^n = r\} = \{\Psi_s^n = r = \Psi_s^{n+1}\} \cap \bigcap_{j<n}\{\Psi_s^j \tilde{\prec} r\} \in \mathcal{F}_r \ \forall s \in S, r \in S,$$

τ_{Ψ_s} is indeed a stopping rule with respect to $(\mathcal{F}_{\Psi_s^n})_{n \in \mathbb{N}_0}$.

For a decision function Ψ_s

$$\mu_{\Psi_s} := \left\{ \begin{array}{ll} \Psi_s^n & \text{on } \{\tau_{\Psi_s} = n\}, n \in \mathbb{N} \\ \hat{\infty} & \text{on } \{\tau_{\Psi_s} = \infty\} \end{array} \right.$$

describes the state at the time of stopping, and

$$Y_{\Psi_s} := \left\{ \begin{array}{ll} Y_r & \text{on } \{\mu_{\Psi_s} = r\}, r \in S_s \\ -\infty & \text{on } \{\mu_{\Psi_s} = \hat{\infty}\} \end{array} \right.$$

gives the corresponding pay-off.

Using the notations

$$\Gamma_s := \left\{ \ \Psi_s : \Psi_s \text{ decision function with starting point } s \in S \text{ s.t. } E\,Y_{\Psi_s} \text{ exists,} \ \right\}$$

the search for an *optimal* decision function means to look for a $\Psi_0^\star \in \Gamma_0$ such that

$$E\,Y_{\Psi_0^\star} = \sup_{\Psi_0 \in \Gamma_0} E\,Y_{\Psi_0}.$$

Again this aim can be achieved, due to an equivalence of the concepts, utilizing our previous results.

(A.21) Lemma

For each decision function Ψ_s, *with starting point* $s \in S$ *there exists a policy* $\pi \in \Pi_s$ *such that*

$$Y_{\alpha(\pi)}(x) = Y_{\Psi_s}(x) \quad \forall x \in \mathcal{X}.$$

Proof: Defining

$$\sigma_0 :\equiv s, \quad \sigma_n := \Psi_s^n, n \in \mathbb{N}$$

and

$$N := \inf\{n \in \mathbb{N}_0 : \sigma_n = \sigma_{n+1}\}$$

we obtain a policy $\pi := ((\sigma_n)_{n \in \mathbb{N}_0}, N)$ with starting value s which satisfies

$$\tau_{\Psi_s}(x) = \inf\{n \in \mathbb{N}_0 : \Psi_s^n(x) = \Psi_s^{n+1}(x)\} = N(x),$$

176

hence $Y_{\alpha(\pi)}(x) = Y_{\Psi_s}(x) \ \forall x \in \mathcal{X}$. Consequently $E\, Y_{\alpha(\pi)} = E\, Y_{\Psi_s}$, i.e. $\pi \in \Pi_s$. $\qquad\square$

Conversely we obtain

(A.22) Lemma

  For each policy $\pi \in \Pi_s$ there exists a decision function $\Psi_s \in \Gamma_s$ such that

$$Y_{\Psi_s}(x) = Y_{\alpha(\pi)}(x) \quad \forall x \in \mathcal{X}.$$

Proof: According to section A.1 we restrict attention to successive policies; for such a policy $\pi = ((\sigma_n)_{n \in \mathbb{N}_0}, N)$

$$\Psi_s(r, x) := \begin{cases} q & \text{if } \sigma_n(x) = r, \sigma_{n+1}(x) = q \text{ for an } n \leq N(x) \\ r & \text{elsewhere} \end{cases}$$

defines a decision function Ψ_s with starting point s. Inductively one obtains that

$$\Psi_s^0 \equiv s \equiv \sigma_0$$

and

$$\Psi_s^n(x) = \Psi_s(\Psi_s^{n-1}(x), x) = \Psi_s(\sigma_{n-1}(x), x) = \sigma_n(x) \ \forall x \in \mathcal{X}, n \in \mathbb{N},$$

in particular $\tau_{\Psi_s}(x) = N(x) \ \forall x \in \mathcal{X}$. Hence it follows that

$$Y_{\Psi_s}(x) = Y_{\alpha(\pi)}(x) \quad \forall x \in \mathcal{X}$$

and $\Psi_s \in \Gamma_s$. $\qquad\square$

As an immediate consequence we obtain that the optimum values for both concepts coincide:

$$\sup_{\Psi_0 \in \Gamma_0} E\, Y_{\Psi_0} = \sup_{\pi \in \Pi_0} E\, Y_{\alpha(\pi)}.$$

Under the general assumption $E(\sup_{s \in S} Y_s^+) < \infty$ lemma (A.7), applied to

$$U_s := \operatorname*{ess\,sup}_{\Psi_s \in \Gamma_s} E(Y_{\Psi_s} | \mathcal{F}_s), \ s \in S,$$

yields

(A.23) Lemma *(Bellman Equation; see [Wa/Wi], p. 963)*

$$U_s = \max\{Y_s, \sup_{r \in D_s} E(U_r | \mathcal{F}_s)\} \ \ P - a.s. \quad \forall s \in S.$$

Proceeding in the same way as for the concept of tactics leads to analogous results about optimal decision functions; moreover, our previous results in the Markov case suggest how to simplify the structure of optimal decision functions under "Markovian" assumptions.

§ A.4 The concept of stopped decision models by Rieder

Considerably more general than the previous ones is the concept of stopped decision models as considered by Rieder [Ri 1], [Ri 2]. In that concept, which may be viewed as a non-stationary dynamic decision model, the transition law can be controlled by the decision maker and the action space may be more complicated than in the previous concepts. On the other hand, Rieder needs rather restrictive (topological) assumptions for his results.

It turns out (for details see [Du] Ch. V) that one can find a natural embedding of problems of optimal sampling (as considered in chapter II) into the framework of stopped decision models: As *state space* the finite sequences of observations are chosen, the *action space* corresponds to $M \cup \{0\}$, the sets of *admissible actions* are

$$\tilde{M}_{(a_1,\ldots,a_{n-1})} \cup \{0\} \quad (\text{if } (a_1,\ldots,a_{n-1}) \in \tilde{A}),$$

the *initial distribution* is the one point measure (Dirac-distribution) in (), and the *transition law* is given by the conditional distribution of the next sub-sample given the "history".

For these special stopped decision models our previous results yield improvements of the assertions of Rieder (due to the fact that we need weaker assumptions). In particular, lemma (2.8), theorem (2.10) and theorem (2.14) generalize the assertions 12.9 and 12.11 of [Ri 1] to special non-compact action spaces and not necessarily upper semi-continuous pay-off functions; the other way round all his results, applied to our special case, are covered by the assertions of Chapter II.

Appendix B: Implementation of the algorithms EV, BF and ILE; Diophantine Approximation[44]

§ B.1 Listing of the modules

The algorithms EV, BF and ILE (see Ch. III, § 4) for computing the OC- and ASC-function of SPPRT's in the iid case were implemented for some classes of discrete distributions (binomial-, Poisson- and negative-binomial-distributions) and SPPRT's of the "onion-skin" type. The programm, written in MODULA, was separated into modules for special tasks (see Ch. III; § 5). We report commented listings of these modules:

The Module AuxiliaryProcedures

```
DEFINITION   MODULE AuxiliaryProcedures;

PROCEDURE max(a,b:INTEGER):INTEGER;

PROCEDURE Mod(a,b:INTEGER):INTEGER;
(* calculates a mod b in the math. sence,
   i.e. a mod b >= 0 *)

PROCEDURE Div(a,b:INTEGER):INTEGER;
(* calculates a div b in the math. sence,
   i.e. a = b * Div(a,b) + Mod(a,b) *)

PROCEDURE IQRoundDown (a,b:INTEGER):INTEGER;
(* calculates RoundDown of the integer quotient a/b *)

PROCEDURE RoundDown(x:REAL):INTEGER;
(* rounds x to the next lower integer number *)

PROCEDURE RoundUp(x:REAL):INTEGER;
(* rounds x to the next greater integer number *)

END AuxiliaryProcedures.
```

---

[44]This appendix is due to Th. Meyerthole

The Module GlobalDefinitions

```
DEFINITION MODULE  GlobalDefinitions;

CONST SampleMax      = 100;     (* maximal number of
                                   additional observations*)
      mMax           = 3200;    (* maximal dimension for the matrix Q *)
      storageMax     = 120000;  (* maximal number of positive
                                   entries  in Q *)

TYPE  CostType = ARRAY [1..SampleMax] OF REAL ;
      (* to store the costfunction *)

      TestDataType= RECORD
                        gamma0,gamma1:REAL;
                        b,t:ARRAY[1..SampleMax]  OF REAL ;
                        indexMax:[1..SampleMax];
                    END;
      (* to store the data of a test in the input representation *)


      TestType    = RECORD
                        g0,g1:INTEGER;
                        Bottom,Top:INTEGER;
                        te:ARRAY [1..mMax] OF CARDINAL ;
                    END;
      (* to store the data of a test in the internal representation *)


      DistributionType = (Poisson,Binomial,NegativeBinomial);
      (* the available distributions *)

PROCEDURE transform(VAR test1:TestDataType; (* in *)
                    VAR test2:TestType      (* in/out *)
                    );
(* transforms the data of a test from the input representation
into the internal representation;
test2.g0 and test2.g1 must be initialized; if necessary the signs
will be changed to guarantee test1.gamma1 * test2.g1 >=0,
but test2.g0/test2.g1 is unchanged *)

END GlobalDefinitions.
```

The Module Densities

```
DEFINITION MODULE  Densities;

(*
exports data structures and procedures to calculate the needed
densities and to calculate gamma1 and gamma0.
*)

FROM GlobalDefinitions IMPORT DistributionType ;

TYPE
    DensityType
        = PROCEDURE(CARDINAL, (* number of convolutions *)
                    REAL,     (* parameter *)
                    INTEGER   (* point with ... *)
                      ):REAL; (* ... probability *)

    DistributionFunctionType
        = PROCEDURE(CARDINAL, (* number of convolutions *)
                    REAL,     (* parameter *)
                    REAL      (* top of interval with ...*)
                      ):REAL;(* ... probability of interval *)

VAR  Density              : DensityType;
                             (* actual Density *)
     DistributionFunction : DistributionFunctionType;
                             (* actual DistributionFunction. *)

PROCEDURE Gamma0(t1,t2:REAL):REAL;

PROCEDURE Gamma1(t1,t2:REAL):REAL;

PROCEDURE SetDistribution(d:DistributionType);
(* to select the desired distribution *)

END Densities.
```

The Module Methods

```
DEFINITION MODULE  Methods;

(*
This module provides the methods BF, EV and ILE.
*)

FROM GlobalDefinitions IMPORT TestType,CostType,
     DistributionType ;

PROCEDURE BF(VAR Test         :TestType;           (* in *)
             VAR Cost         :CostType;           (* in *)
             p                :REAL;               (* in *)
             d                :DistributionType;(* in *)
             start            :INTEGER;            (* in *)
             delta            :REAL;               (* in *)
             VAR ASC,OC,PF     :REAL;               (* out *)
             VAR rest         :REAL ;              (* out *)
             VAR iterat       :CARDINAL            (* out *)
            );
PROCEDURE EV(VAR Test         :TestType;           (* in *)
             VAR Cost         :CostType;           (* in *)
             p                :REAL;               (* in *)
             d                :DistributionType;(* in *)
             start            :INTEGER;            (* in *)
             delta            :REAL;               (* in *)
             VAR EVused        :BOOLEAN;            (* out *)
             VAR ASC,OC,PF     :REAL;               (* out *)
             VAR rest         :REAL ;              (* out *)
             VAR iterat       :CARDINAL            (* out *)
            );
PROCEDURE ILE(VAR Test        :TestType;           (* in *)
              VAR Cost        :CostType;           (* in *)
              p               :REAL;               (* in *)
              d               :DistributionType;  (* in *)
              start           :INTEGER;            (* in *)
              delta           :REAL;               (* in *)
              newinit         :BOOLEAN;            (* in *)
              VAR ASC,OC       :REAL;               (* out *)
              VarOption       :BOOLEAN;            (* in *)
              VAR Var         :REAL;               (* out *)
              VAR iASC,iOC     :CARDINAL;           (* out *)
              VAR iVar        :CARDINAL            (* out *)
             );

END Methods.
```

```modula-2
IMPLEMENTATION MODULE  Methods;

(*
This module provides the methods BF, EV and ILE.
*)

FROM GlobalDefinitions IMPORT TestType,CostType,mMax,
                              DistributionType,storageMax;
FROM MathLib0 IMPORT real ;
IMPORT Densities;
FROM Densities IMPORT Density,DistributionFunction,SetDistribution;
FROM AuxiliaryProcedures IMPORT max,Mod,Div,IQRoundDown;
FROM InOut IMPORT WriteString,WriteLn,Read,CloseOutput;


TYPE VectorType = ARRAY [1..mMax] OF REAL ;

VAR m               : CARDINAL;  (* dimension of the matrix Q *)
    pp,ppold        :VectorType; (* to store the distribution
                                    on the states
                                    of the Markov chain *)
    C,q1,q2         :VectorType;
    i               :CARDINAL;
    EVstop,BFstop:BOOLEAN;
    ASChelp,OChelp,PFhelp:REAL;
    lambda:   REAL;
    sumpp,sumppold,factor:REAL;

(*****************************************************************)
(*                                                             *)
(*              module for matrix-operations                   *)
(*                                                             *)
(*****************************************************************)

MODULE Markov;
(*
exports all procedures to handle the matrix Q
*)

FROM Densities IMPORT Density,DistributionFunction;

IMPORT mMax,storageMax,TestType,m,Div,Mod,max,VectorType,
       qii,WriteLn,WriteString,CloseOutput,Read;
EXPORT QtMatMult,InitQ,SSum,Initqii;
```

```modula
TYPE Matrix = RECORD
                  column :ARRAY [1..mMax] OF
                                RECORD  offset:INTEGER;
                                        length:INTEGER;
                                        row:LONGCARD;
                                END ;
                  values :ARRAY [1..storageMax]  OF REAL ;
              END;


VAR Q:Matrix;

PROCEDURE  error;
VAR ch:CHAR;
BEGIN
CloseOutput;
WriteString(' storage exceeded ');
WriteLn;
WriteString(' press a key ');
Read(ch);
HALT;
END error;

PROCEDURE InitQ(VAR Test:TestType;
                    p    :REAL    );
VAR i:INTEGER;
    k:CARDINAL;
    z:LONGCARD;
BEGIN
z:=1;
FOR i:=1 TO m  DO
 WITH Q.column[i] DO
  WITH Test DO
    offset:=Mod(-(1-i+g0*INTEGER(te[i])), ABS(g1)) +1;
    length:=Div(INTEGER(m)-offset , ABS(g1))+1;
    row:=z;
    z:=z+LONG(CARDINAL(max(0,length)));
    IF z>storageMax THEN
          error;
    END;
    FOR k:=1 TO length DO
      Q.values[(row+LONG(k))]:=
        Density(te[i],p,
          Div(offset-i+g0*INTEGER(te[i])+(INTEGER(k)-1)*ABS(g1),g1));
    END;
  END;
 END;
END;
END InitQ;
```

184

```
PROCEDURE QtMatMult(VAR Test :TestType;(* in *)
                    VAR ppold,         (* in  *)
                        pp             (* out *)
                          :VectorType );


(*
calculates in pp the product of Q transposed applied to ppold;
for that purpose pp and ppold must be different
*)
VAR i,k:CARDINAL;
BEGIN
FOR i:=1 TO m DO
   pp[i]:=0.0;
END;
FOR i:=1 TO m DO
  WITH Q.column[i] DO
    WITH Test DO
      FOR k:=1 TO length DO
        pp[offset+INTEGER(k-1)*ABS(g1)]:=
                  pp[offset+INTEGER(k-1)*ABS(g1)]
                  +ppold[i]*Q.values[(row+LONG(k))];
      END;
    END;
  END;
END;
END QtMatMult;



PROCEDURE SSum(VAR Test:TestType;(* in *)
               VAR x:VectorType; (* in *)
                   i:CARDINAL       (* in *)
                 ):REAL ;


(*
calculates q[i,1]*x[1]+ ... +q[i,m]*x[m]   - q[i,i]*x[i]
*)
VAR sum:REAL;
    k:CARDINAL;
```

```
BEGIN (* SSum *)
sum:=0.0;
WITH Q.column[i]  DO
  WITH Test  DO
    FOR k:=1 TO length DO
      sum:= sum+Q.values[row+LONG(k)]*x[offset+INTEGER(k-1)*ABS(g1)];
    END;
    sum:=sum-qii[i]*x[i];
  END;
END;
RETURN sum;
END SSum;


PROCEDURE Initqii(VAR Test:TestType);
(* initializes qii with the diagonal elements of Q *)
VAR i,k:CARDINAL;
BEGIN
WITH Test DO
  FOR i:=1 TO m DO
    qii[i]:=0.0;
    WITH Q.column[i] DO
      IF Mod(INTEGER(i)-offset,ABS(g1))=0 THEN
        k:=Div(INTEGER(i)-offset,ABS(g1))+1;
        IF (1<=k)&(INTEGER(k)<=length) THEN
            qii[i]:=Q.values[row+LONG(k)];
        END;
      END;
    END;
  END;
END;
END Initqii;

END Markov;
```

```
(***********************************************************)
(*                                                       *)
(*               small auxiliary-procedures              *)
(*                                                       *)
(***********************************************************)

PROCEDURE copy(VAR a,    (* in  *)
                   b     (* out *)
                     :VectorType);
VAR i:CARDINAL;
BEGIN
FOR i:=1 TO m DO b[i]:=a[i] END;
END copy;

PROCEDURE SMultpp(VAR a:VectorType(* in *)):REAL;
(* calculates pp[1]*a[1]+ ... +pp[m]*a[m] *)
VAR i: CARDINAL;
    help:REAL;
BEGIN
help:=0.0;
FOR i:=1 TO m DO help:=help+pp[i]*a[i] END;
RETURN help;
END SMultpp;

PROCEDURE sum(VAR a:VectorType (* in *)):REAL;
(* calculates a[1]+ ... + a[m] *)
VAR i: CARDINAL;
    help:REAL;
BEGIN
help:=0.0;
FOR i:=1 TO m DO help:=help+a[i] END;
RETURN help;
END sum;

PROCEDURE Initq1(VAR Test:TestType;p:REAL);
VAR i:CARDINAL;
BEGIN
WITH Test DO
 IF (g1 > 0) THEN
   FOR i:=1 TO m DO
     q1[i]:=1.0-
       DistributionFunction(te[i],p,
         real(IQRoundDown(INTEGER(m-i)+g0*INTEGER(te[i]),g1)));
   END
```

```
  ELSE
    FOR i:=1 TO m DO
      q1[i]:=
        DistributionFunction(te[i],p,
          real(IQRoundDown(INTEGER(m+1-i)+g0*INTEGER(te[i]),g1)));
    END
  END;
END;
END Initq1;

PROCEDURE Initq2(VAR Test:TestType;p:REAL);
VAR i:INTEGER;
BEGIN
WITH Test DO
 IF (g1>0)THEN
   FOR i:=1 TO m DO
    q2[i]:=
        DistributionFunction(te[i],p,
            real(IQRoundDown((-i)+g0*INTEGER(te[i]),g1)));
   END;
 ELSE
   FOR i:=1 TO m DO
    q2[i]:=1.0-
      DistributionFunction(te[i],p,
            real(IQRoundDown(g0*INTEGER(te[i])-i+1,g1)));
   END;
 END;
END
END Initq2;

PROCEDURE Initpp(VAR Test:TestType;start:INTEGER),
VAR i: CARDINAL;
BEGIN
FOR i:=1 TO m DO pp[i]:=0.0 END;
pp[start-Test.Bottom]:=1.0;
END Initpp;

PROCEDURE InitC(VAR Test:TestType;VAR Cost:CostType);
VAR i:CARDINAL;
BEGIN
FOR i:=1 TO m DO C[i]:=Cost[Test.te[i]] END;
END InitC;
```

188

```
(**************************************************************)
(*                                                          *)
(*                     BF - method                          *)
(*                                                          *)
(**************************************************************)

PROCEDURE BF(VAR Test          :TestType;         (* in *)
             VAR Cost          :CostType;         (* in *)
             p                 :REAL;             (* in *)
             d                 :DistributionType;(* in *)
             start             :INTEGER;          (* in *)
             delta             :REAL;             (* in *)
             VAR ASC,OC,PF     :REAL;             (* out *)
             VAR rest          :REAL ;            (* out *)
             VAR iterat        :CARDINAL          (* out *)
            );
PROCEDURE Init;
BEGIN
SetDistribution(d);
m:=Test.Top-Test.Bottom-1;
InitQ(Test,p);
Initpp(Test,start);
Initq1(Test,p);
Initq2(Test,p);
InitC(Test,Cost);
ASC:=SMultpp(C);
OC:=SMultpp(q2);
PF:=SMultpp(q1);
END Init;

BEGIN (* BF *)
Init;
i:=0;
REPEAT
    INC(i);
    copy(pp,ppold);
    QtMatMult(Test,ppold,pp);
    ASC:=ASC+SMultpp(C);
    OC :=OC +SMultpp(q2);
    PF :=PF +SMultpp(q1);
    rest:=sum(pp);
    BFstop:=(rest<delta)OR((1.0-(OC+PF))<delta);
UNTIL BFstop;
iterat:=i;
END BF;
```

```modula-2
(**********************************************************)
(*                                                        *)
(*                    EV - method                         *)
(*                                                        *)
(**********************************************************)

PROCEDURE EV(VAR Test        :TestType;         (* in *)
             VAR Cost        :CostType;         (* in *)
             p               :REAL;             (* in *)
             d               :DistributionType;(* in *)
             start           :INTEGER;          (* in *)
             delta           :REAL;             (* in *)
             VAR EVused      :BOOLEAN;          (* out *)
             VAR ASC,OC,PF   :REAL;             (* out *)
             VAR rest        :REAL ;            (* out *)
             VAR iterat      :CARDINAL          (* out *)
             );

PROCEDURE Init;
BEGIN
SetDistribution(d);
m:=Test.Top-Test.Bottom-1;
InitQ(Test,p);
Initpp(Test,start);
Initq1(Test,p);
Initq2(Test,p);
InitC(Test,Cost);
ASC:=SMultpp(C);
OC:=SMultpp(q2);
PF:=SMultpp(q1);
END Init;

CONST MinIterat=3;

BEGIN (* EV *)
Init;
FOR i:=1 TO MinIterat DO
        copy(pp,ppold);
        QtMatMult(Test,ppold,pp);
        ASC:=ASC+SMultpp(C);
        OC :=OC +SMultpp(q2);
        PF :=PF +SMultpp(q1);
END;
```

```
i:=MinIterat;
EVstop:=FALSE;
BFstop:=FALSE;
sumpp:=sum(pp);
factor:=1.0;
REPEAT
    INC(i);
    copy(pp,ppold);
    QtMatMult(Test,ppold,pp);
    sumppold:=sumpp;
    sumpp:=sum(pp);
    IF sumppold # 0.0 THEN lambda:=sumpp/sumppold;END;
    IF lambda # 1.0 THEN factor:=1.0/(1.0-lambda);END;
    ASChelp:=SMultpp(C);
    OChelp :=SMultpp(q2);
    PFhelp :=SMultpp(q1);
    EVstop:=(ABS(1.0-(OC+OChelp*factor)-(PF+PFhelp*factor))<delta);
    IF EVstop THEN ASC:=ASC+ASChelp*factor;
                   OC :=OC +OChelp*factor;
                   PF :=PF +PFhelp*factor;
              ELSE ASC:=ASC+ASChelp;
                   OC :=OC +OChelp;
                   PF :=PF +PFhelp;
    END;
    rest:=sumpp;
    BFstop:=(rest<delta)OR((1.0-(OC+PF))<delta);
UNTIL EVstop OR BFstop;
EVused:=EVstop;
iterat:=i;
END EV;

(*************************************************************)
(*                                                         *)
(*                    ILE - method                         *)
(*                                                         *)
(*************************************************************)


VAR LC,LC2,Lq2,     (* to store the LE-systems
                       approximative solutions *)
    qii,            (* to store the
                       diagonal elements of Q *)
    C2:VectorType;  (* to calculate the variance *)
```

```
PROCEDURE ILE(VAR Test          :TestType;            (* in *)
              VAR Cost          :CostType;            (* in *)
              p                 :REAL;                (* in *)
              d                 :DistributionType;    (* in *)
              start             :INTEGER;             (* in *)
              delta             :REAL;                (* in *)
              newinit           :BOOLEAN;             (* in *)
              VAR ASC,OC        :REAL;                (* out *)
              VarOption         :BOOLEAN;             (* in *)
              VAR Var           :REAL;                (* out *)
              VAR iASC,iOC      :CARDINAL;            (* out *)
              VAR iVar          :CARDINAL             (* out *)
             );

VAR iterat:CARDINAL;
    ASC2:REAL;
    maxdiff,help:REAL;

PROCEDURE InitLC;
BEGIN
copy(C,LC);
END InitLC;

PROCEDURE InitLC2;
BEGIN
copy(C2,LC2);
END InitLC2;

PROCEDURE InitLq2;
BEGIN
copy(q2,Lq2);
END InitLq2;

PROCEDURE InitC2;
VAR i:CARDINAL;
    sum:REAL;
BEGIN
FOR i:=1 TO m DO
  C2[i]:=C[i]*C[i]+2.0*C[i]*(SSum(Test,LC,i)+qii[i]*LC[i]);
END;
END InitC2;
```

```
PROCEDURE Init;
BEGIN
SetDistribution(d);
m:=Test.Top-Test.Bottom-1;
InitQ(Test,p);
Initq2(Test,p);
InitC(Test,Cost);
Initqii(Test);
IF newinit THEN
  InitLC;
  InitLq2;
END;
END Init;


BEGIN (* ILE *)
Init;

iterat:=0;
REPEAT
  INC(iterat);
  maxdiff:=0.0;
  FOR i:=1 TO m DO
    help :=(C[i]+SSum(Test,LC,i))/(1.0-qii[i]);
    IF ABS(LC[i]-help) > maxdiff THEN
            maxdiff:=ABS(LC[i]-help)
    END;
    LC[i]:=help;
  END;
UNTIL maxdiff < delta;
iASC:=iterat;
ASC:= LC[start-Test.Bottom];

IF VarOption  THEN
  InitC2;
  IF newinit THEN
      InitLC2;
  END;
```

```
    iterat:=0;
    REPEAT
      INC(iterat);
      maxdiff:=0.0;
      FOR i:=1 TO m DO
        help  :=(C2[i]+SSum(Test,LC2,i))/(1.0-qii[i]);
        IF ABS(LC2[i]-help)>maxdiff THEN
            maxdiff:=ABS(LC2[i]-help)
        END;
        LC2[i]:=help;
      END;
    UNTIL maxdiff<delta;
    iVar:=iterat;
    ASC2:= LC2[start-Test.Bottom];
    Var:=ASC2-(ASC*ASC);
  END;

  iterat:=0;
  REPEAT
    INC(iterat);
    maxdiff:=0.0;
    FOR i:=1 TO m DO
      help  :=(q2[i]+SSum(Test,Lq2,i))/(1.0-qii[i]);
      IF ABS(help-Lq2[i])>maxdiff THEN
          maxdiff:=ABS(help-Lq2[i])
      END;
      Lq2[i]:=help;
    END;
  UNTIL maxdiff < delta;
  iOC:=iterat;
  OC := Lq2[start-Test.Bottom];

END TLE;


END Methods.
```

194

§ B.2 Diophantine approximation

For the algorithms of Ch. III; § 4 we need "good" approximations of real numbers $a \in \mathbb{R}$ by rational numbers p/q, $p \in \mathbf{Z}$, $q \in \mathbb{N}$, with "small" denominator q. The key for this problem is the *continued fraction expansion*. One can write a in the form

$$a = a_0 + \cfrac{1}{a_1 + \cfrac{1}{a_2 + \cdots}}$$

where $a_0 = \lfloor a \rfloor$ and $a_1, a_2, \ldots$ are natural numbers which can be computed recursively by

$$b_0 := a \quad ; \quad a_0 := \lfloor b_0 \rfloor,$$
$$b_{k+1} := \frac{1}{b_k - a_k} \quad ; \quad a_{k+1} := \lfloor b_{k+1} \rfloor.$$

This computation is stopped if $b_k = a_k = 0$. In this case a is exactly described by a finite expansion (and hence rational). The number

$$\frac{p_k}{q_k} = a_0 + \cfrac{1}{a_1 + \cfrac{1}{\ddots \atop a_{k-1} + \frac{1}{a_k}}}$$

where p_k, q_k are normed s.t. $q_k > 0$, is called the kth *convergent* of a.

Obviously

$$p_0 = a_0, \ q_0 = 1; \ p_1 = a_0 a_1 + 1, \ q_1 = a_1;$$

moreover

(B.1) Theorem: *The convergents p_k/q_k of a fulfill*

i.) $\left| a - \frac{p_k}{q_k} \right| \leq \frac{1}{q_k q_{k+1}}$.

ii.) $(q_k)_{k \in \mathbb{N}_0}$ *grows exponentially fast.*

iii.) $\lim_{k \to \infty} p_k/q_k = a$.

iv.) $\begin{aligned} p_{k+1} &= a_{k+1} p_k + p_{k-1}, \\ q_{k+1} &= a_{k+1} q_k + q_{k-1}. \end{aligned}$

v.) $p_{k+1} q_k - p_k q_{k+1} = (-1)^k$.

vi.) $\frac{p_0}{q_0} \leq \frac{p_2}{q_2} \leq \frac{p_4}{q_4} \leq \ldots a \ldots \leq \frac{p_3}{q_3} \leq \frac{p_1}{q_1}$.

For a proof of i.)-v.) comp. [Lo], p. 10/11; for part vi.) see [Sdt], Lemma I.4A. Property iii.) guarantees that each real number may be approximated arbitrarily precise by continued fractions. Using property iv.), the convergents p_k/q_k may be computed successively.

In the sequel we discuss some problems of rational approximation. Without loss of generality we assume $a \geq 0$; then all values a_k, p_k, q_k, $k \in \mathbb{N}_0$, are non-negative.

(B.2) Problems *(see [Lo], p. 11/12)*
   Let $a \geq 0$, $\varepsilon > 0$ and $q_{max} \in \mathbb{N}$ be given.

1. *Find $p, q \in \mathbb{N}_0$ such that $|a - \frac{p}{q}| < \varepsilon$ and q is minimal, i.e.*

$$q = \min\{q \in \mathbb{N} : \exists\, p \in \mathbb{N}_0 : |a - \frac{p}{q}| < \varepsilon\}.$$

2. *Find $p, q \in \mathbb{N}_0$ such that $q \leq q_{max}$ and $|a - \frac{p}{q}|$ is as small as possible, i.e.*

$$|a - \frac{p}{q}| = \min\{|a - \frac{p}{q}| : p, q \in \mathbb{N}_0, q \leq q_{max}\}.$$

3. *Find $p, q \in \mathbb{N}_0$ such that $q \leq q_{max}$, $|a - \frac{p}{q}| < \varepsilon$, and q minimal if possible, elsewhere $|a - \frac{p}{q}|$ as small as possible, i.e. for*

$$
\begin{aligned}
Z &:= \{q \leq q_{max} : \exists\, p \in \mathbb{N}_0 : |a - \frac{p}{q}| < \varepsilon\} \\
q &= \min Z, \text{ if } Z \neq \emptyset \\
|a - \frac{p}{q}| &= \min\{|a - \frac{p}{q}| : p, q \in \mathbb{N}_0, q \leq q_{max}\}, \text{ if } Z = \emptyset.
\end{aligned}
$$

The solutions of these problems make use of

(B.3) *Let $a, b, c, d \in \mathbb{N}$ such that $ad > bc$; then*

$$\frac{a}{b} - \frac{c}{d} \geq \frac{1}{bd}$$

which follows from $ad \geq bc + 1$ by multiplying with $\frac{1}{bd}$:

(B.4) Theorem *(solution of problem 1; see [Lo], p.12)*
   *For given $a \geq 0$ and $\varepsilon > 0$ let $k \in \mathbb{N}_0$ be minimal s.t. $|a - p_k/q_k| < \varepsilon$.
   For $k = 0$*

$$p_0/q_0 = a_0/1$$

solves problem 1.
For $k = 1$ let $i > 0$ be the smallest integer such that $|a - (i\, a_0 + 1)/i| < \varepsilon$. Then

$$\frac{i\, a_0 + 1}{i}$$

solves problem 1.
For $k > 1$ let $i > 0$ be the smallest integer such that

$$|a - \frac{p_{k-2} + i\, p_{k-1}}{q_{k-2} + i\, q_{k-1}}| < \varepsilon.$$

Then

$$\frac{p_{k-2} + i\, p_{k-1}}{q_{k-2} + i\, q_{k-1}};$$

solves problem 1.

Proof: The case $k = 0$ is obvious. For $k = 1$ we have

$$\frac{p_1}{q_1} = \frac{a_0 a_1 + 1}{a_1} \in (a - \varepsilon, a + \varepsilon)$$

by assumption; hence there exists an i with the desired property and $0 < i \leq a_1$. Let now $s \leq a_1$, $s \in \mathrm{IN}$ and $r \in \mathbf{Z}$ s.t. $|a - r/s| < \varepsilon$; then this property is also fulfilled for that $r' \in \mathbf{Z}$ which minimizes

$$\left| a - \frac{r}{s} \right| = \frac{|s\, a - r|}{|s|}.$$

But this r' is either sa_0 or $sa_0 + 1$. In the first case one obtains $r'/s = a_0$ which contradicts the choice of k; in the second case $r'/s = (sa_0 + 1)/s$, i.e. the solution has to be of this form. The definition of i yields the smallest possible denominator.

For $k \geq 2$, k odd (the other case is treated analogously) one obtains due to the choice of k

$$\frac{p_{k-1}}{q_{k-1}} \leq a - \varepsilon < a \leq \frac{p_k}{q_k} < a + \varepsilon \leq \frac{p_{k-2}}{q_{k-2}};$$

where according to (B.1)(iv)

$$\frac{p_k}{q_k} = \frac{p_{k-2} + a_k p_{k-1}}{q_{k-2} + a_k q_{k-1}}.$$

Hence the definition of i makes sense, and $0 < i \leq a_k$. Furthermore

$$\frac{p_{k-2} + (i - 1)p_{k-1}}{q_{k-2} + (i - 1)q_{k-1}} \geq \frac{p_{k-2} + i p_{k-1}}{q_{k-2} + i q_{k-1}}$$

since k is odd (see (B.1)v.)). Because of the choice of i this yields

$$a + \varepsilon \leq \frac{p_{k-2} + (i - 1)p_{k-1}}{q_{k-2} + (i - 1)q_{k-1}}.$$

Hence we obtain for arbitrary $r/s \in (a - \varepsilon, a + \varepsilon)$

$$\frac{p_{k-1}}{q_{k-1}} < \frac{r}{s} < \frac{p_{k-2} + (i - 1)p_{k-1}}{q_{k-2} + (i - 1)q_{k-1}}.$$

It remains to show that $s \geq q_{k-2} + i q_{k-1}$. But applying (B.3) on the last two inequalities leads to

$$\frac{r}{s} - \frac{p_{k-1}}{q_{k-1}} \geq \frac{1}{s q_{k-1}}$$

and

$$\frac{p_{k-2} + (i - 1)p_{k-1}}{q_{k-2} + (i - 1)q_{k-1}} - \frac{r}{s} \geq \frac{1}{s(q_{k-2} + (i - 1)q_{k-1})}.$$

Adding up one obtains

$$\frac{1}{sq_{k-1}} + \frac{1}{s(q_{k-2} + (i-1)q_{k-1})}$$

$$\leq \frac{p_{k-2} + (i-1)p_{k-1}}{q_{k-2} + (i-1)q_{k-1}} - \frac{p_{k-1}}{q_{k-1}}$$

$$= \frac{-(-1)^{k-2}}{q_{k-1}(q_{k-2} + (i-1)q_{k-1})} \text{ according to (B.1)v.);}$$

since k is odd this yields

$$s \geq q_{k-2} + (i-1)q_{k-1} + q_{k-1} = q_{k-2} + iq_{k-1} \qquad \square$$

(B.5) Theorem *(solution of problem 2; see [Lo], p.11/12)*

If $p_k/q_k = a$ for a k such that $q_k \leq q_{max}$, this rational number solves the problem. Let elsewhere p_k/q_k be the first convergent of a such that $q_k > q_{max}$ and choose the largest $i \in \mathrm{IN}$ such that $q_{k-2} + iq_{k-1} \leq q_{max}$. Then either

$$\frac{p_{k-2} + ip_{k-1}}{q_{k-2} + iq_{k-1}} \qquad or \qquad \frac{p_{k-1}}{q_{k-1}}$$

solves problem 2.

For a proof see [Lo], p. 11.

(B.6) Theorem *(solution of problem 3)*

Let k be the smallest integer such that

$$|a - \frac{p_k}{q_k}| < \varepsilon \quad or \quad q_k > q_{max}.$$

If $|a - p_k/q_k| < \varepsilon$, let p/q be the solution of problem 1 (according to (B.4)); for $q \leq q_{max}$ this is also a solution for problem 3. If $|a - p_k/q_k| \geq \varepsilon$ or $q > q_{max}$, a solution of problem 2 (according to (B.5)) is also a solution of problem 3.

Proof: It is sufficient to show that the set Z (defined in (B.2)3.) is empty in the case

$$|a - p_k/q_k| \geq \varepsilon \quad \text{and} \quad q_k > q_{max}$$

as well as in the case

$$|a - p_k/q_k| < \varepsilon \quad \text{and} \quad q > q_{max}.$$

In the first case this is an immediate consequence of (B.4) since the minimal denominator is $\geq q_k > q_{max}$. In the second case the denominator q is $> q_{max}$ by assumption. $\qquad \square$

These propositions allow to find, for a given real number, a diophantine approximation
– with minimal denominator for prescribed accuracy
– with optimal accuracy for a prescribed upper bound of the denominator
– with prescribed upper bound for the denominator and either minimal denominator under an accuracy restriction or minimal deviation from that bound if the restriction

cannot be fulfilled.

This is of importance since the denominator of the diophantine approximation for γ_0/γ_1 in Ch. III, § 4 determines the size of the matrix $\overline{Q}$ (see (3.18)). Efficiency arguments give reason to keep, for given capacity restrictions and certain accuracy requirements, the size of that matrix as small as possible.

For our example (3.34) (see also p. 11) we considered the simple alternatives

$$p_1 = 0.90016837105 \quad , \quad p_2 = 0.95;$$

the reason for the somewhat curious choice of p_1 was that we arrive for $\varepsilon = 10^{-4}$ with $k = 5, i = 1$ at the "pleasant" diophantine approximation

$$g_0/g_1 = 77/83$$

for $\gamma_0/\gamma_1 = 0.927710843$ (see (3.13)) (yielding an actual accuracy $< 5.10^{-8}$).

Appendix C: References, Bibliography

On the one hand we give references to the literature used in these lecture notes, on the other hand our aim is to arrange a bibliography of papers relevant to sequentially planned decision procedures. For the sake of an easy access to these references we add short characterizations of the respective topics according to the following list of categories:

GS   G̲roup s̲equential methods (with fixed group sizes)
SP   Sequential s̲ampling p̲lans (variable group sizes)
MI   M̲ultistage i̲nspection plans (variable sample sizes)
PO   Optimal stopping of processes with p̲artially o̲rdered index sets
ME   M̲ulti-stage e̲stimators
SA   S̲equential a̲nalysis (general)
OS   O̲ptimal s̲topping (general)
⋆   Additional papers/books cited in the text.

Our goal is to give an (as complete as possible) bibliography for the categories SP, MI and PO; for the other points merely some important papers and useful books are mentioned.

[Ab 1]   Abel, V.: Kostenoptimale mehrstufige Bayes-Inspektionspläne.
Ph.D. thesis, Heidelberg 1978   MI

[Ab 2]   -   :   Some essentially complete classes for Bayesian multistage sampling inspection.
Commun. Statist. A9(14), (1980), 1483-1489.   MI

[Ab 3]   -   :   Multistage sampling procedures based on prior distributions and costs for infinite populations.
Meth. Oper. Res. 39(1981), 7-15. comp. also
Meth. Oper. Res. 41(1981), 113-115.   MI

   -   :   On the efficiency of Bayesian multistage sampling inspection procedures.
Meth. Oper. Res. 44(1981), 445-455.   MI

Armitage, P.: *Sequential Medical Trials.*
Blackwell, Oxford 1975; 2nd Ed.   SA

[Ba]   Bauer, H.: *Wahrscheinlichkeitstheorie und Grundzüge der Maßtheorie.*
de Gruyter, Berlin, 3rd Ed. 1978.   ⋆

[Be]   Berger, J.: *Statistical Decision Theory. Foundations, Concepts and Methods.*
Springer, New York etc., 1980 (2nd Ed. 1985)   ⋆

[Bek]   Berk, R.H.: Locally most powerful sequential tests.
Ann. of Statist. 3(1975), 373-381   SA

[B/D]   Bickel, P., K. Doksum: *Mathematical Statistics. Basic Ideas and Selected Topics.*
Holden-Day, San Francisco 1977   ⋆

Blum, J.E., J. Rosenblatt: On multistage estimation.
    Ann. Math. Statist. 34(1963), 1452-1458 — ME

[B/P] Brown, L.,R. Purves: Measurable selection of extrema.
    Ann. Statist. 1(1973), 902-912 — $\star$

Chernoff, H.: Sequential design of experiments.
    Ann. Math. Statist. 30(1959), 755-770 — (SP)

Chernoff, H., S.N. Ray: A Bayes sequential sampling inspection plan.
    Ann. Math. Statist. 36(1965), 1387-1407 — MI

Chow, Y.S.: Martingales in a σ-finite measure space indexed by directed sets.
    Trans. Amer. Math. Soc. 97(1960), 254-285 — PO

[C/R] Chow, Y.S., H. Robbins: On the asymptotic theory of fixed width confidence intervals for the mean.
    Ann. Math. Statist. 36(1965), 457-462. — SA

[C/R/S] Chow, Y.S., H. Robbins, D. Siegmund: *Great Expectations: The Theory of Optimal Stopping.*
    Houghton Mifflin, Boston 1971 — OS

[C/M] Cox, D., H. Miller: *The Theory of Stochastic Processes.*
    Methuen & Co, London, 1965 — $\star$

Cressie, N., P. Morgan: The VPRT: Optimal sequential and non-sequential testing. In: Statistical Decision Theory and Related Topics IV (Ed. S.S. Gupta and J.O. Berger); Springer, N.Y. 1988; 107-118 — SP

[Da] Dantzig, G.: On the non-existence of tests of "Student's" hypothesis having power functions independent of σ.
    Ann. Math. Statist. 11(1940), 186-192 — $\star$

[DeG] DeGroot, M.:*Optimal Statistical Decisions.*
    McGraw-Hill, New York 1970 — $\star$

DeMets, D.L., J.H. Ware: Group sequential methods for clinical trials with a one-sided hypothesis.
    Biometrika 67(1980), 651-660 — GS

  - : Asymmetric group sequential boundaries for monitoring clinical trials.
    Biometrika 69(1982), 661-663 — GS

[D/U 1] Dieter, U.,M. Unger: Sequentielle Analysis: Genaue Werte für die Bernoulli-Verteilung. Österr. Zeitschrift für Statistik und Informatik 17(1987), 27-47 — SA

[D/U 2] - : Sequential analysis: Exact values for the Bernoulli distribution. Contribution to Stochastics (Ed. W. Sendler), Physica, Heidelberg 1987; 50-58 — SA

[D/R] Dodge, H., H. Romig: *Sampling Inspections Tables – Single and Double Sampling.*
    Wiley, New York 1959 — MI

[Do]   Doob, J.: *Stochastic Processes.*
        Wiley, New York, 1953   ⋆

[Du]   Duscha, G.:   Optimale sequentiell geplante Verfahren bei partiell
        geordneten Indexmengen.
        Ph.D. thesis, Münster 1990   PO/SP

        Eger, K.-H.: Zur Berechnung der Operationscharakteristik
        und des mittleren Stichprobenumfangs bei Mehrfach-
        stichprobenplänen zur Attributprüfung.
        Wiss. Z. der TH Karl-Marx-Stadt 19(1977), 467-476   GS

[Eg]   -   :   *Sequential Tests.*
        Teubner, Leipzig 1982   SA

[E]   Ehrenfeld, S.: On group sequential sampling.
        Technometrics 14(1972), 167-174.   SP

        Elfring, G.L., J.R. Schultz: Group sequential designs for clinical trials.
        Biometrics 29(1973), 471-477   GS

[E/M]   Enkawa, T., M. Mori: Exact solutions for OC- and ASN-functions of
        Poisson sequential probability ratio tests.
        Rep. of Statist. Appl. Research, Juse 32(1985), 1-16.   SA

[Fe]   Ferguson, Th.: *Mathematical Statistics. A Decision Theoretic Approach.*
        Academic Press, New York 1967   ⋆

        Flohrer, J.: Sequentielle Schätzung bei gruppierter Stichprobennahme.
        Wiss. Z. der TH Karl-Marx-Stadt 19(1977), 477-489   SP

[Gh]   Ghosh, B.K.: *Sequential Tests of Statistical Hypotheses.*
        Addison-Wesley, Reading, 1970   SA

        Girshick, M.A., S. Karlin, H.L. Royden: Multistage statistical decision
        procedures.
        Ann. Math. Statist. 28(1957), 111-125   SA

        Gould, A.L., V.J. Pecore: Group sequential methods for clinical trials
        allowing early acceptance of H_0 and incorporating costs.
        Biometrica 69(1982), 75-80   GS

[Ha]   Haggstrom, G.W.: Optimal stopping and experimental design.
        Ann. Math. Statist. 37(1966), 7-29.   SP

[Had]   Hald, A.: *Statistical Theory of Sampling Inspection by Attributes.*
        Academic Press, London 1981   MI

[Hal]   Hall, P.:   Asymptotic theory of triple sampling for sequential
        estimation of a mean.
        Ann. Statist. 9(1981), 1229-1238   SP

        Harenbrock, M.: Optional Sampling Theoreme für Submartingale
        mit partiell geordneten Indexmengen.
        Ph.D. thesis, Münster 1990   PO

202

-,   N. Schmitz: Optional sampling of submartingales with scanned index sets.
Journal Theor. Probability 5(1992), 309-326   PO

Havalec, L., V. Scheiber, F.X. Wohlzogen: Gruppierungspläne für sequentielle Testverfahren.
Int. J. Clin. Pharmacol. 3(1971), 342-345   GS

Hayre, L.S.: Adaptive multistage sampling procedures for attributes.
Frontiers in Statistical Quality Control 2 (Ed. Lenz et al.), 40-50; Physica, Würzburg, 1984   MI

[Hay]   -   : Group sequential sampling with variable group sizes.
J. Roy. Statist. Soc. Ser. B 47(1985), 90-97.   SP

Heckendorf, H.: *Grundlagen der sequentiellen Statistik.*
Teubner, Leipzig 1982   SA

-      : Sufficiency in sequential experimental design problems.
In: Sequential Methods in Statistics. Banach Center Publ. Vol. 16, 181-192 Polish Scientific Publishers 1985   SA

Hürzeler, H.E.: Quasimartingales on partially ordered sets.
J. Multivar. Anal. 14(1984), 34-73.   PO

-      : The optional sampling theorem for processes indexed by a partially ordered set.
Ann. Probab. 13(1985), 1224-1235   PO

[Ir]   Irle, A.:   *Sequentialanalyse. Optimale sequentielle Tests.*
Teubner, Stuttgart 1990   OS/SA

Jennison, Ch.: Efficient group sequential tests with unpredictable group sizes.
Biometrika 74(1987), 155-165   GS(SP)

[J/T]   Jennison, Ch., B. Turnbull: Statistical approaches to interim monitoring of medical trials.
Statist. Science 5(1990), 299-317   GS/SP

[K/W]   Kiefer, J., L. Weiss: Some properties of generalized sequential probability ratio tests.
Ann. Math. Statist. 28(1957), 57-74   SA

[K/S]   Krengel, U., L. Sucheston: Stopping rules and tactics for processes indexed by a directed set.
J. Multivariate Anal. 11(1981), 199-229   PO

Kurtz, T.G.: The optional sampling theorem for martingales indexed by directed sets.
Ann. Probab. 8(1980), 675-681   PO

Lan, K.K., D.L. DeMets: Discrete sequential boundaries for clinical trials.
Biometrika 70(1983), 659-663   GS(SP)

Lawler, G.F., R.J. Vanderbei: Markov strategies for optimal control
problems indexed by a partially ordered set.
Ann. Probab. 11(1983), 642-647   PO

[Le]   Lehmann, E.: *Testing Statistical Hypotheses.*
Wiley, New York 1959 (2nd Ed. 1986)   ⋆

[Lor]   Lorden, G.:   Structure of sequential tests minimizing an expected
sample size.
Z. Wahrsch.theorie verw. Geb. 51(1980), 291-302   SA
- :   Asymptotic efficiency of three-stage hypothesis tests.
Ann. Statist. 11(1983), 129-140   SP

[Lo]   Lovacs, L.:   *An Algorithmic Theory of Numbers, Graphs and Convexity.*
SIAM, Philadelphia 1986   ⋆

[Lü]   Lübbert, J.:   Optimale sequentielle Selektionsprozeduren und
optimale sequentielle Stichprobenpläne.
Ph.D. thesis, Münster 1988   SP

[M/V]   Mandelbaum, A., R.J. Vanderbei: Optimal stopping and super-
martingales over partially ordered sets.
Z. Wahrscheinlichkeitsth. 57(1981), 253-264   PO

[MA]   Martinsek, A.T.: Multistage estimation: Optimal and asymptotically
optimal policies.
Sequential Analysis 5(1986), 1-17   ME
Menges, G., M. Behara: Das Bayes'sche Risiko bei sequentiellen
Stichprobenentscheidungen.
Stat. Hefte 3(1962), 39-61   SP

[Me]   Meyerthole, Th.: Optimale sequentiell geplante Entscheidungsverfahren
IV: Sequentiell geplante Tests; sequentiell geplante SPRT's.
Angew. Math. und Informatik 6/90-S, Münster 1990   SP
Millet, A.:   On randomized tactics and optimal stopping in the plane.
Ann. Prob. 13(1985), 946-965   PO
Mukhopadhyay, N.: A note on three-stage and sequential point
estimation procedures for a normal mean.
Sequential Analysis 4(1985), 311-319   ME

[Pf]   Pfanzagl,J.:   Sampling procedures based on prior distributions
and costs.
Technometrics 5(1963), 47-61   MI
Pocock, S.J.: Group sequential methods in the design and analysis
of clinical trials.
Biometrika 64(1977), 191-199   GS
- :   Interim analyses for randomized clinical trials:
The group sequential approach.
Biometrics 38(1982), 153-162   GS

[Ri 1]   Rieder, U.:   Bayessche Dynamische Entscheidungs- und Stoppmodelle.
Ph.D. thesis, Hamburg 1972   [OS/SA]

[Ri 2]   - :   Bayesian Dynamic Programming.
Adv. Appl. Prob. 7(1975), 330-348   (SA)

[R/W]   -, P. Wentges: On Bayesian group sequential sampling procedures.
Ann. of Oper. Research 32(1991), 189-203   SP

Schmitz, N.: Elemente einer Theorie der optimal geplanten
sequentiellen Entscheidungsverfahren.
Preprint Münster, 1985   SP

-       : From optimal stopping to optimal sampling.
In: Statistik, Informatik und Ökonometrie (Ed. W. Janko);
Springer, N.Y. 1988; 272-289   SP

[Sch]   -   : Optimale sequentiell geplante Entscheidungsverfahren.
Teil 1: Theorie.
Skripten zur Math. Statistik Nr. 18; Münster 1989   SP

-       : Wald-Wolfowitz optimality of sequentially planned
tests - remarks and conjectures.
Ann. of Oper. Research 32(1991), 205-213   SP

Schneider, H., K.-H. Waldmann: Cost optimal multistage sampling
plans. Frontiers in Statistical Quality Control 2
(Ed. Lenz et al.); 32-39, Physica, Würzburg 1984   MI

[Schü] Schüler, W.: Multistage sampling procedures based on prior
distributions and costs.
Ann. Math. Statist. 38(1967), 464-470   MI

Schüler, W., J. Pfanzagl: The efficiency of sequential sampling plans
based on prior distributions and costs.
Technometrics 12(1970), 299-310   MI

[Se]   Seneta, E.: *Non-Negative Matrices.*
Allen & Unwin, London, 1973   ⋆

Sicking, P.: *Bayessche Mehrstufige Qualitätskontrolle.*
P. Lang, Frankfurt, 1990   MI

[Si]   Siegmund, D.: Some problems in the theory of optimal stopping
rules.
Ann. Math. Statist. 38(1967), 1627-1640   OS

[S/E]   Spahn, M., S. Ehrenfeld: Optimal and suboptimal procedures in
group sequential sampling.
Nav. Res. Log. Q. 21(1974), 53-68   VG

[St]   Stein, Ch.: A two-sample test for a linear hypothesis whose
power is independent of the variance.
Ann. Math. Statist. 16(1945), 243-258   ME

[St 2] -   : A note on cumulative sums.
Ann. Math. Statist. 17(1946), 498-499   SA

[Wa]   Wald, A.: *Sequential Analysis.*
J. Wiley, N.Y. 1947   SA

-       : *Statistical Decision Functions.*
J. Wiley, N.Y. 1950   SA

[W/W] -; Wolfowitz, J.: Optimum character of the sequential probability
                ratio test.
                Ann. Math. Statist. 19(1948), 326-339   SA

[Wal 1] Waldmann, K.H.: Multistage Bayesian acceptance sampling: Optimality
                of a (z, c^-, c^+)-sampling in case of a Polya prior distribution.
                Stat. & Dec. 3(1985), 167-186   MI

       -        : Kostenoptimale Bayessche Stichprobenpläne für die
                Eingangs- und Endkontrolle von Warenlieferungen.
                Oper. Res. Proc. 1984; 573-580 Springer, Berlin u.a. 1985 MI

       -        : Computational aspects in multistage Bayesian
                acceptance sampling.
                Naval Res. Log. Quarterly 33(1986), 399-412   MI

       -        : Sufficient conditions for optimality of a $a(z, c^-, c^+)$-
                sampling plan in multistage Bayesian acceptance sampling.
                OR-Spektrum 9(1987), 23-31

[Wa/Wi]Washburn, R.B., A.S. Willsky: Optional sampling of submartingales
                indexed by partially ordered sets.
                Ann. Prob. 9(1981), 957-970   PO

[We]   Wetherill, G.B.: *Sequential Methods in Statistics.*
                Chapman and Hall 1975   SA

        Whitehead, J.: *The Design and Analysis of Sequential Clinical Trials.*
                E. Horwood, Chichester 1983   SA

        Whittle, P.: Some general results in sequential design (with discussion).
                J. Roy. Statist. Soc. Ser. B. 27(1965), 371-394   SP

[Wi 1]   Witting, H.: *Mathematische Statistik I*
                Teubner, Stuttgart 1985   ⋆

[Wi 2]   -; Müller-Funk, U.: *Mathematische Statistik II*
                Teubner, Stuttgart (forthcoming)   ⋆

[W/R]   Woodall, W., M. Reynolds: A discrete Markov chain representation
                of the sequential probability ratio test.
                Commun. Statist. Sequent. Analysis 2(1983), 27-44   SA

        Ziegler, W.J.: Zum Problem der Optimum-Eigenschaften von SPR-Tests.
                Contributions to Applied Statistics 1976; 257-262;
                Birkhäuser, Basel   SP

Vol. 1: R.A. Fisher: An Appreciation. Edited by S.E. Fienberg and D.V. Hinkley. XI, 208 pages, 1980.

Vol. 2: Mathematical Statistics and Probability Theory. Proceedings 1978. Edited by W. Klonecki, A. Kozek, and J. Rosinski. XXIV, 373 pages, 1980.

Vol. 3: B.D. Spencer, Benefit-Cost Analysis of Data Used to Allocate Funds. VIII, 296 pages, 1980.

Vol. 4: E.A. van Doorn, Stochastic Monotonicity and Queueing Applications of Birth-Death Processes. VI, 118 pages, 1981.

Vol. 5: T. Rolski, Stationary Random Processes Associated with Point Processes. VI, 139 pages, 1981.

Vol. 6: S.S. Gupta and D.-Y. Huang, Multiple Statistical Decision Theory: Recent Developments. VIII, 104 pages, 1981.

Vol. 7: M. Akahira and K. Takeuchi, Asymptotic Efficiency of Statistical Estimators. VIII, 242 pages, 1981.

Vol. 8: The First Pannonian Symposium on Mathematical Statistics. Edited by P. Révész, L. Schmetterer, and V.M. Zolotarev. VI, 308 pages, 1981.

Vol. 9: B. Jørgensen, Statistical Properties of the Generalized Inverse Gaussian Distribution. VI, 188 pages, 1981.

Vol. 10: A.A. McIntosh, Fitting Linear Models: An Application of Conjugate Gradient Algorithms. VI, 200 pages, 1982.

Vol. 11: D.F. Nicholls and B.G. Quinn, Random Coefficient Autoregressive Models: An Introduction. V, 154 pages, 1982.

Vol. 12: M. Jacobsen, Statistical Analysis of Counting Processes. VII, 226 pages, 1982.

Vol. 13: J. Pfanzagl (with the assistance of W. Wefelmeyer), Contributions to a General Asymptotic Statistical Theory. VII, 315 pages, 1982.

Vol. 14: GLIM 82: Proceedings of the International Conference on Generalised Linear Models. Edited by R. Gilchrist. V, 188 pages, 1982.

Vol. 15: K.R.W. Brewer and M. Hanif, Sampling with Unequal Probabilities. IX, 164 pages, 1983.

Vol. 16: Specifying Statistical Models: From Parametric to Non-Parametric, Using Bayesian or Non-Bayesian Approaches. Edited by J.P. Florens, M. Mouchart, J.P. Raoult, L. Simar, and A.F.M. Smith, XI, 204 pages, 1983.

Vol. 17: I.V. Basawa and D.J. Scott, Asymptotic Optimal Inference for Non-Ergodic Models. IX, 170 pages, 1983.

Vol. 18: W. Britton, Conjugate Duality and the Exponential Fourier Spectrum. V, 226 pages, 1983.

Vol. 19: L. Femholz, von Mises Calculus For Statistical Functionals. VIII, 124 pages, 1983.

Vol. 20: Mathematical Learning Models — Theory and Algorithms: Proceedings of a Conference. Edited by U. Herkenrath, D. Kalin, W. Vogel. XIV, 226 pages, 1983.

Vol. 21: H. Tong, Threshold Models in Non-linear Time Series Analysis. X, 323 pages, 1983.

Vol. 22: S. Johansen, Functional Relations, Random Coefficients and Nonlinear Regression with Application to Kinetic Data, VIII, 126 pages, 1984.

Vol. 23: D.G. Saphire, Estimation of Victimization Prevalence Using Data from the National Crime Survey. V, 165 pages, 1984.

Vol. 24: T.S. Rao, M.M. Gabr, An Introduction to Bispectral Analysis and Bilinear Time Series Models. VIII, 280 pages, 1984.

Vol. 25: Time Series Analysis of Irregularly Observed Data. Proceedings, 1983. Edited by E. Parzen. VII, 363 pages, 1984.

Vol. 26: Robust and Nonlinear Time Series Analysis. Proceedings, 1983. Edited by J. Franke, W. Härdle and D. Martin. IX, 286 pages, 1984.

Vol. 27: A. Janssen, H. Milbrodt, H. Strasser, Infinitely Divisible Statistical Experiments. VI, 163 pages, 1985.

Vol. 28: S. Amari, Differential-Geometrical Methods in Statistics. V, 290 pages, 1985.

Vol. 29: Statistics in Ornithology. Edited by B.J.T. Morgan and P.M. North. XXV, 418 pages, 1985.

Vol 30: J. Grandell, Stochastic Models of Air Pollutant Concentration. V, 110 pages, 1985.

Vol. 31: J. Pfanzagl, Asymptotic Expansions for General Statistical Models. VII, 505 pages, 1985.

Vol. 32: Generalized Linear Models. Proceedings, 1985. Edited by R. Gilchrist, B. Francis and J. Whittaker. VI, 178 pages, 1985.

Vol. 33: M. Csörgo, S. Csörgo, L. Horváth, An Asymptotic Theory for Empirical Reliability and Concentration Processes. V, 171 pages, 1986.

Vol. 34: D.E. Critchlow, Metric Methods for Analyzing Partially Ranked Data. X, 216 pages, 1985.

Vol. 35: Linear Statistical Inference. Proceedings, 1984. Edited by T. Calinski and W. Klonecki. VI, 318 pages, 1985.

Vol. 36: B. Matérn, Spatial Variation. Second Edition. 151 pages, 1986.

Vol. 37: Advances in Order Restricted Statistical Inference. Proceedings, 1985. Edited by R. Dykstra, T. Robertson and F.T. Wright. VIII, 295 pages, 1986.

Vol. 38: Survey Research Designs: Towards a Better Understanding of Their Costs and Benefits. Edited by R.W. Pearson and R.F. Boruch. V, 129 pages, 1986.

Vol. 39: J.D. Malley, Optimal Unbiased Estimation of Variance Components. IX, 146 pages, 1986.

Vol. 40: H.R. Lerche, Boundary Crossing of Brownian Motion. V, 142 pages, 1986.

Vol. 41: F. Baccelli, P. Brémaud, Palm Probabilities and Stationary Queues. VII, 106 pages, 1987.

Vol. 42: S. Kullback, J.C. Keegel, J.H. Kullback, Topics in Statistical Information Theory. IX, 158 pages, 1987.

Vol. 43: B.C. Arnold, Majorization and the Lorenz Order: A Brief Introduction. VI, 122 pages, 1987.

Vol. 44: D.L. McLeish, Christopher G. Small, The Theory and Applications of Statistical Inference Functions. VI, 124 pages, 1987.

Vol. 45: J.K. Ghosh (Ed.), Statistical Information and Likelihood. 384 pages, 1988.

Vol. 46: H.-G. Müller, Nonparametric Regression Analysis of Longitudinal Data. VI, 199 pages, 1988.

Vol. 47: A.J. Getson, F.C. Hsuan, {2}-Inverses and Their Statistical Application. VIII, 110 pages, 1988.

Vol. 48: G.L. Bretthorst, Bayesian Spectrum Analysis and Parameter Estimation. XII, 209 pages, 1988.

Vol. 49: S.L. Lauritzen, Extremal Families and Systems of Sufficient Statistics. XV, 268 pages, 1988.

Vol. 50: O.E. Barndorff-Nielsen, Parametric Statistical Models and Likelihood. VII, 276 pages, 1988.

Vol. 51: J. Hüsler, R.-D. Reiss (Eds.), Extreme Value Theory. Proceedings, 1987. X, 279 pages, 1989.

Vol. 52: P.K. Goel, T. Ramalingam, The Matching Methodology: Some Statistical Properties. VIII, 152 pages, 1989.

Vol. 53: B.C. Arnold, N. Balakrishnan, Relations, Bounds and Approximations for Order Statistics. IX, 173 pages, 1989.

Vol. 54: K.R. Shah, B.K. Sinha, Theory of Optimal Designs. VIII, 171 pages, 1989.

Vol. 55: L. McDonald, B. Manly, J. Lockwood, J. Logan (Eds.), Estimation and Analysis of Insect Populations. Proceedings, 1988. XIV, 492 pages, 1989.

Vol. 56: J.K. Lindsey, The Analysis of Categorical Data Using GLIM. V, 168 pages, 1989.

Vol. 57: A. Decarli, B.J. Francis, R. Gilchrist, G.U.H. Seeber (Eds.), Statistical Modelling. Proceedings, 1989. IX, 343 pages, 1989.

Vol. 58: O.E. Barndorff-Nielsen, P. Blæsild, P.S. Eriksen, Decomposition and Invariance of Measures, and Statistical Transformation Models. V, 147 pages, 1989.

Vol. 59: S. Gupta, R. Mukerjee, A Calculus for Factorial Arrangements. VI, 126 pages, 1989.

Vol. 60: L. Györfi, W. Härdle, P. Sarda, Ph. Vieu, Nonparametric Curve Estimation from Time Series. VIII, 153 pages, 1989.

Vol. 61: J. Breckling, The Analysis of Directional Time Series: Applications to Wind Speed and Direction. VIII, 238 pages, 1989.

Vol. 62: J.C. Akkerboom, Testing Problems with Linear or Angular Inequality Constraints. XII, 291 pages, 1990.

Vol. 63: J. Pfanzagl, Estimation in Semiparametric Models: Some Recent Developments. III, 112 pages, 1990.

Vol. 64: S. Gabler, Minimax Solutions in Sampling from Finite Populations. V, 132 pages, 1990.

Vol. 65: A. Janssen, D.M. Mason, Non-Standard Rank Tests. VI, 252 pages, 1990.

Vol. 66: T. Wright, Exact Confidence Bounds when Sampling from Small Finite Universes. XVI, 431 pages, 1991.

Vol. 67: M.A. Tanner, Tools for Statistical Inference: Observed Data and Data Augmentation Methods. VI, 110 pages, 1991.

Vol. 68: M. Taniguchi, Higher Order Asymptotic Theory for Time Series Analysis. VIII, 160 pages, 1991.

Vol. 69: N.J.D. Nagelkerke, Maximum Likelihood Estimation of Functional Relationships. V, 110 pages, 1992.

Vol. 70: K. Iida, Studies on the Optimal Search Plan. VIII, 130 pages, 1992.

Vol. 71: E.M.R.A. Engel, A Road to Randomness in Physical Systems. IX, 155 pages, 1992.

Vol. 72: J.K. Lindsey, The Analysis of Stochastic Processes using GLIM. VI, 294 pages, 1992.

Vol. 73: B.C. Arnold, E. Castillo, J.-M. Sarabia, Conditionally Specified Distributions. XIII, 151 pages, 1992.

Vol. 74: P. Barone, A. Frigessi, M. Piccioni, Stochastic Models, Statistical Methods, and Algorithms in Image Analysis. VI, 258 pages, 1992.

Vol. 75: P.K. Goel, N.S. Iyengar (Eds.), Bayesian Analysis in Statistics and Econometrics. XI, 410 pages, 1992.

Vol. 76: L. Bondesson, Generalized Gamma Convolutions and Related Classes of Distributions and Densities. VIII, 173 pages, 1992.

Vol. 77: E. Mammen, When Does Bootstrap Work? Asymptotic Results and Simulations. VI, 196 pages, 1992.

Vol. 78: L. Fahrmeir, B. Francis, R. Gilchrist, G. Tutz (Eds.), Advances in GLIM and Statistical Modelling: Proceedings of the GLIM92 Conference and the 7th International Workshop on Statistical Modelling, Munich, 13-17 July 1992. IX, 225 pages, 1992.

Vol. 79: N. Schmitz, Optimal Sequentially Planned Decision Procedures. XII, 209 pages, 1992.